NATIONAL BIODIVERSITY PLANNING

Guidelines Based on Early Experiences Around the World

World Resources Institute
in cooperation with
United Nations Environment Programme
The World Conservation Union (IUCN)

1995

This report represents a timely, technical contribution to the efforts of countries around the world as they engage in the implementation of the Convention on Biological Diversity. Its sponsors take responsibility for choosing and focusing the study topics and guaranteeing its authors and researchers freedom of inquiry. The report's authors have also solicited and responded to the guidance of leaders and experts from countries currently working on this topic, and to peer review workshops and correspondence. The views of the authors expressed in this report do not necessarily reflect those of the United Nations Environment Programme (UNEP),The World Conservation Union IUCN), World Resources Institute (WRI), or the participating countries. Unless otherwise stated, all the interpretations and findings set forth here, and analyses of the national case studies, are the sole responsibility of the authors.

Citation: Miller, Kenton. R. and Steven M. Lanou. 1995. National Biodiversity Planning: Guidelines Based on Early Experiences Around the World. World Resources Institute, United Nations Environment Programme and The World Conservation Union. Washington D. C.; Nairobi; Gland, Switzerland.

ISBN: 1-56973-025-3
Library of Congress Catalog Card No. 95-061095

Additional copies of this report can be obtained by contacting WRI Publications, P.O. Box 4852, Hampden Station, Baltimore, MD, 21211, USA; tel: 1-800-822-0504 or 410-516-6963

Printed on recycled paper

Kathleen Courrier
Publications Director

Brooks Belford
Marketing Manager

Hyacinth Billings
Production Manager

Bruce Cabarle, Walter Reid
Cover Photos

Pamela Reznick
Cover Design

Gary Ridley
Book Design

Contents

Foreword vii
Acknowledgments ix
Executive Summary xi

CHAPTER I

Introduction to this Guide 1
Introduction 3
Background 5
Method of this Study 6
How to Use this Guide 7

CHAPTER II

Relationships among Biodiversity-Planning Tools and Between Strategic Biodiversity Planning and Other Planning Efforts . . . 9
Introduction 11
Relationships among Biodiversity-Planning Tools 11
Relationship between Strategic Biodiversity Planning and Other Planning Efforts 12
Integration of Biodiversity Planning into Existing Plans . . . . .
Integration of Biodiversity Planning into Multiple Sectors . . . . .

CHAPTER III

Lessons Learned from Early Experiences 21
Introduction 23
Ten Guiding Principles 23
Obstacles to Biodiversity Planning 27
Ingredients for Success 28
Other Lessons from the Country Cases 28

CHAPTER IV

A Framework for National Biodiversity Planning: The Convention on Biological Diversity 35
Introduction 37
Objectives of the Convention 37
Checklist of Action Points 37
Provisions of the Convention in Brief 38

CHAPTER V
An Illustrative Biodiversity-Planning Process: The Basic Steps 43
Introduction 45
Illustrative Biodiversity-Planning Steps in Brief 45
Illustrative Biodiversity-Planning Steps in Detail 45

ANNEX A
Profiles in National Biodiversity Planning 85
Australia 87
Canada 90
The Philippines 92
Germany 95
Norway 97
Indonesia 99
United Kingdom 102
Vietnam 104
The Netherlands 106
China 109
Chile 111
Mexico 114
Poland 117
Kenya 119
Egypt 121
Costa Rica 123
South Pacific Biodiversity Conservation Program 125

ANNEX B
Background Information on Intergovernmental Decisions and Corresponding Activities that Provide Overall Mandate for this Study 129

ANNEX C
Criteria for Identifying Countries with Advanced Experience in Strategic Biodiversity Planning 135

ANNEX D
Pro Forma Questionnaire for Case Study Reports 139

ANNEX E
Contributors to the Publication 143

ANNEX F
Contents of the Convention on Biodiversity to be Addressed in National Strategies and Action Plans 149

References 159

LIST OF BOXES

BOX 1: *Key Documentation that Should be Made Available to Support Biodiversity Planning*.....5
BOX 2: *Strategies for Sustainable Development Handbook Series*.....6
BOX 3: *The Biodiversity Planning Process: Expanding Participation and Benefits*.....13
BOX 4: *Basic Steps for Biodiversity Planning and its Relationship to the Planning Tools: A Cyclical and Adaptive Process*.....14
BOX 5: *Relationship Among Biodiversity Tools, Planning Steps, and Outcomes*.....15
BOX 6: *Linkages of Selected International, National, and Regional Environment and Development Strategies*.....17
BOX 7: *Example of a Framework for Sectoral Plans*.....18
BOX 8: *Characteristics that Distinguish the Capacity among Countries for Biodiversity Planning and Implementation*.....24
BOX 9: *Ten Guiding Principles for Biodiversity Planning*.....25
BOX 10: *Abbreviated and Selected List of Provisions of the Biodiversity Convention to which Action can be Addressed*.....39
BOX 11: *Basic Steps in an Illustrative Biodiversity Planning Process*.....46
BOX 12: *Nurturing Political Commitment*.....47
BOX 13: *Sample Work Sheet. Record of Responsibilities for Implementing the Biodiversity Plan*.....69
BOX 14: *Some Potential Sources of International Support for Biodiversity Strategies and Action Plans*.....70
BOX 15: *Essential Elements to be Included in a Monitoring and Evaluation Program*.....78
BOX 16: *Illustrative Table of Contents for a National Biodiversity Strategy Report*.....82
BOX 17: *Illustrative Table of Contents for a National Biodiversity Action Plan Report*.....83

ILLUSTRATIVE BOXES

BOX 1: *Step 1. Getting Organized: Steering Committee and Working Groups*.....49
BOX 2: *Step 2. Assessment: Inventory*.....51
BOX 3: *Assessment: Financial Resources and Unmet Needs for Conservation*.....52
BOX 4: *Assessment: Value of Biodiversity*.....52
BOX 5: *Assessment: Economic Benefits of Biodiversity*.....54
BOX 6: *Assessment: Costs of Biodiversity Conservation*.....55
BOX 7: *Assessment: Management Costs for Parks and Protected Areas*.....56
BOX 8: *Step 3. Developing a Strategy: National Vision, Guiding Principles, and Goals*.....59
BOX 9: *Developing a Strategy: Goals*.....60
BOX 10: *Developing a Strategy: Objectives*.....60
BOX 11: *Developing a Strategy: Principles*.....61
BOX 12: *Developing a Strategy: Operational Objectives and Targets for Key Habitats and Species*.....62
BOX 13: *Developing a Strategy: Components of a Strategy or Action Plan*.....63
BOX 14: *Developing a Strategy: Components of a Biodiversity Program*.....64
BOX 15: *Developing a Strategy: Key Elements for Initial Action*.....65
BOX 16: *Developing a Strategy: Criteria for Determining Biodiversity Significance*.....65
BOX 17: *Developing a Strategy: Criteria for Determining Priorities for Actions*.....66
BOX 18: *Developing a Strategy: Traditional Knowledge*.....67
BOX 19: *Developing a Strategy: Indigenous Peoples' Participation*.....68
BOX 20: *Step 4. Developing an Action Plan: Priority Actions and Time Frames*.....72
BOX 21: *Developing an Action Plan: Economic Incentives and Penalties that Promote Biodiversity Conservation*... 74
BOX 22: *Developing an Action Plan: Bioregional Management*..... 75
BOX 23: *Developing an Action Plan: Access to Genetic Resources*.....76
BOX 24: *Developing an Action Plan: Project Formulation*..... 77
BOX 25: *Steps 5 and 6. Implementation and Monitoring: Objectives for Monitoring*.....79
BOX 26: *Step 6. Monitoring: Components of a Monitoring System*..... 80
BOX 27: *Monitoring: Linkages between Data, Information, Monitoring and Development of a Biodiversity Strategy and Action Plan*.....81

Foreword

Awareness of the Earth's dwindling biotic wealth spread far and wide during the three years leading up to the 1992 Earth Summit in Rio de Janeiro. Once governments everywhere began recognizing how invaluable—and how endangered—biodiversity is, they embraced the idea that something had to be done to improve the ways we use biological resources to benefit both the generations now living and those to come.

This shared sense of urgency led 156 nations and the European Union to sign the legally binding Convention on Biological Diversity during the Rio conference.

Many others have signed since. As they ratify the convention, governments accept responsibility to safeguard and comprehend the profusion of species, genetic materials, habitats, and ecosystems that make up the natural world. They agree to foster development that uses biological resources sustainably. They agree to recognize each nation's sovereignty over the biodiversity found in its territory. What's more, they agree to correct the imbalance between who benefits from biodiversity protection and who pays—committing to find equitable ways to share biodiversity's monetary and non-monetary values, to spur technology cooperation, and to establish mechanisms to finance investments in maintaining the diversity of life on Earth.

The national biodiversity strategies and action plans called for in Article 6 are key vehicles for implementing the Convention. The process of preparing such plans can not only help each country articulate its own priorities for domestic actions and for international cooperation, it can also strengthen the capacity of its people and institutions to address the full array of Convention mandates.

As leaders of the three institutions that spearheaded this study, we are pleased to offer these guidelines and country profiles to governments, community leaders, nongovernmental organizations, indigenous groups, and leaders of business and industry—all of whom must play design and implementation roles if national strategies and action plans are to live up to their promise. The backbone of this independent analysis is the pioneering work carried out by eighteen countries—some developed, some developing, some in transition from centrally-planned to market economies, and some small island states—all of which freely shared their experiences with creating national strategies and action plans by writing case studies, talking with our researchers, and taking part in a peer review workshop.

The paramount lesson learned so far is how crucial it is to engage everyone with a stake in the outcome in the preparation of national strategies and action plans, so that "biodiversity planners" become "biodiversity implementors." Shaping policies and plans that are in both a nation's interest and nature's interest requires input from all sectors of government and society, just as carrying out such plans demands widespread cooperation.

Like all efforts to implement the biodiversity convention, these guidelines are necessarily preliminary. Such questions as how to prepare project proposals for financial support, ways to control access to genetic resources, and modes of technological cooperation still await decisions by the Convention's Conference of the Parties. Other practical aspects, such as how to weave biodiversity provisions throughout government policies, will remain experimental until there is widespread experience to learn from. Scientists and field practitioners are continually refining their techniques for assessing biotic resources. Thus, we anticipate that countries will adapt, test, and revise these guidelines as they develop their national strategies and plans—and that new guidelines will eventually evolve that take account of all this added experience and knowledge.

Since 1988, the World Resources Institute, the United Nations Environment Programme, and The World Conservation Union have been partners in a joint international biodiversity conservation program, beginning with the process that produced the 1992 *Global Biodiversity Strategy,* which is available in eight languages. Our cooperative effort will continue to seek ways to broaden the international constituency and the debate on biodiversity matters through published research on key issues, workshops, briefings, and such mechanisms as the ongoing Global Biodiversity Forum.

Jonathan Lash
President
World Resources Institute

Elizabeth Dowdeswell
Executive Director
United Nations Environment Programme

David McDowell
Director General
The World Conservation Union

Acknowledgments

This report was the product of a cooperative effort headed by Kenton R. Miller and Steven M. Lanou of the World Resources Institute, in collaboration with the United Nations Environment Programme (UNEP) and the World Conservation Union (IUCN). The report has benefitted tremendously from the advice of Hamdallah Zedan, Paul Chabeda, and Feargul Duff at UNEP and from the able guidance and essential contributions of Allen Putney, Jeffrey McNeely, Jeremy Carew-Reid, and Martha Rojas at IUCN.

These guidelines are the result of extensive inquiry and discussion with many countries and individuals from all the world's regions. The case studies prepared by partner countries provide the central research basis of the report. Hence, a special word of appreciation is owed to Marc Auer, Richard Bagine, Charles Barber, Roger Bendall, Esam Ahmed Elbadry, Wang Enmin, Rodrigo Gámez, John Herity, Mohamed Isahakia, Jorge Larson, Jeffrey McNeely, Josephine Mummery, Consuelo Muñoz, Iosefatu Reti, Bart Romijin, Peter Schei, Gudrun Schneider, Jorge Soberón, Wang Sung, Marcel Vernooy, and Andrzej Weigle. More details concerning the case study contributors can be found in Annex E.

The report also owes much to those who commented on the three draft manuscripts. A list of reviewers who submitted written comments is in Annex E. WRI colleagues Walt Reid, Nels Johnson, Chip Barber, Laura Lee Dooley, and Walter Arensberg provided essential insights into the whole research and drafting process.

The final document was revised by a peer review workshop held at Williamsburg, Virginia, that brought together individuals from all regions who have actual experience in biodiversity planning in their own countries or on associated biodiversity-planning research. The document gained much from their collective experience and guidance. A complete list of the workshop participants can be found in Annex E.

Finally, the hard work and dedication of Lisa Sullivan and Donna Dwiggins who helped organize the Williamsburg Workshop, assisted in the preparation of the report, and kept the whole process on track, deserve a special thank you.

Executive Summary

One hundred and fifty-six governments made a statement at the Earth Summit at Rio de Janeiro in 1992: They are prepared to accept responsibility for conserving the full diversity of plant, animal, and microbial life in their countries, to begin using biological resources sustainably, and to seek the equitable sharing of benefits from biodiversity. Rich nations committed themselves to providing technological and financial resources to help poor countries build the capacity to save, manage, and employ their biotic wealth.

The commitment made at Rio had its roots in the United Nations Environment Programme Governing Council's declaration of the need for concerted international action for effectively conserving the world's biodiversity. Among the earliest steps taken by countries before Rio was initiating country studies—the systematic assessment of their biodiversity. Later, within the text of the Convention, the Parties agreed to formulate national strategies and action plans and to integrate biodiversity activities into all relevant sectors. This guide offers a method that "biodiversity planners" can use to initiate a national biodiversity planning process that builds upon country studies and other planning efforts.

At the invitation and with the strong support of the United Nations Environment Programme (UNEP), the World Resources Institute (WRI) initiated this study in 1992. Also, the World Conservation Union (IUCN) contributed its considerable experience with national strategy work in developing countries and its extensive capability in areas related to biodiversity.

Eighteen countries joined with WRI, UNEP, and IUCN to examine their own experiences with biodiversity planning. Although much of this experience is recent, these developing and developed nations, along with those with economies in transition and the small island states, have already learned a great deal that will be of value to those who follow. Seventeen countries provided a written case study on their work—the basis for later interviews by WRI and IUCN project staff. Most participated directly in the review of early drafts and in the peer review workshop held at Williamsburg, Virginia, in September 1994.

Drawing from this early experience, an illustrative seven-step biodiversity planning process is spelled out here as a guide to those willing to make choices and get them implemented.

1. Getting organized—establish a focal point in government, get an adequate high-level mandate, form a partnership with governmental agencies, nongovernmental organizations (NGOs), indigenous peoples, community leaders, and business and industry, and obtain adequate funds.

2. *Assessment (country study)*—gather and evaluate information on the status and trends of the nation's biodiversity and biological resources, laws, policies, organizations, programs, budgets, and human capacity; select preliminary goals and objectives; identify gaps between desired and current situations; review options to close gaps; and estimate costs, benefits, and unmet needs.

3. *Developing a strategy*—determine goals and operational objectives; analyze and select specific measures to close the gaps identified in the assessment; hold further consultations and dialogue until consensus is reached on acceptable targets and mechanisms for action; and identify the potential roles of stakeholder groups.

4. *Developing a plan of action*—determine which public and private organizations and groups will implement which activities denoted in the strategy, in which location or region, by what means, and with which people, institutions, facilities, and funds, and set a time table for action.

5. *Implementation*—launch activities and policies in practical ways so that partners take charge of particular elements of the plan and biodiversity planners become "biodiversity implementors;" in other words, individuals from the key ministries, NGOs, communities, indigenous groups, business, and industries, each with self, group, or business interests and commitment move forward to seek results from their plans and action.

6. *Monitoring and evaluation*—observe and measure the impact of the plan on the economy, ecosystems, and social indicators; note changes in laws and policies, behavioral responses, conservation improvement, sustainability, and enhanced equity; and note changes in capacity and investment.

7. *Reporting*—prepare reports for important constituencies; such documents can include country studies, national strategies, action plans, reports to the Convention, and reports to the country's chief executive and general public.

Biodiversity planning is an open-ended process that develops continuously as further information and experience is gained. The process is cyclical, with the same steps repeated round after round. It is adaptive because participants learn from past experience about shifts in nature and society, and it also involves multiple stakeholders and sectors. A partnership is needed among all those committed to making choices and taking action, from all parts of society, and from all sectors of government and the economy.

The preliminary lessons learned from the countries that cooperated in this study provide considerable guidance to those countries seeking to undertake a similar process. The array of institutional, scientific, legal, and policy obstacles encountered by these countries were met with actions that show promise and innovation. Among the most potent factors for facilitating biodiversity planning and action are solid political will and commitment by the highest levels of government.

The process of developing guidelines for biodiversity planning will necessarily be an iterative one. Several key components of the planning process await policy guidance from the Parties to the Convention on action. Such topics as project formulation, criteria and priorities for access to the Convention's funding mechanism, technology cooperation, and access to genetic resources can be expected to enter the deliberations and workplans of the Conference of the Parties. But since any attempt to offer guidelines on these topics right now would be speculative and without empirical grounding, users are urged to consider how Parties can clarify policies and criteria. Parties and their subsidiary scientific and technical bodies can set the stage for future documentation. Future guidelines can be expected to address these and other critical topics, again drawing from fieldwork.

Experience already shows how all the countries studied have successfully used an open, participatory approach. Transitional economies have found themselves with outstanding human, scientific, technological, and management capability, but often without clear mandates and political and institutional commitment. Distinctively, small island states have found considerable value in regional collaboration, sharing well-experienced personnel and focusing attention on priorities of common interest. The developing countries may have only limited experience with comprehensive planning and limited funding for planning and implementation, but their expertise in relevant technical fields is typically noteworthy. In contrast, developed countries often find themselves dealing with complex public institutions with overlapping jurisdictions and with no interest in changing how they operate.

It is hoped that this guide will help agency officials, NGOs, communities, indigenous people, and business leaders orient their initial efforts in biodiversity planning. The guide should also help in capacity-building workshops and team planning. No doubt, it will be supplemented by national-level guides as soon as countries develop their own methods and formulate materials for dealing with their peculiar issues and opportunities.

Chapter One • Introduction to this Guide

In this Chapter:

- *Introduction*
- *Background*
- *Method of this Study*
- *How to Use this Guide*

Introduction to this Guide

INTRODUCTION

The future of life on Earth captured worldwide attention at the Earth Summit in Rio de Janeiro in 1992 when 155 nation states and the European Union signed the Convention on Biological Diversity (UNEP, 1992). This act signaled their intention to join with other nations to form a global cooperative to protect habitats, species, and genes, to shift to sustainable modes of resource use, and to make the necessary policy, economic, and managerial adjustments to guarantee that the benefits to be gained from forest and range ecosystems, soils, agricultural production, wildlife management, fisheries, and genetic resources are equitably shared across local, regional, national, and global societies.

Nations also adopted a comprehensive global work plan for national actions and international cooperation for sustainable development and global environmental protection well into the 21st century. Named *Agenda 21*, the plan contains 40 chapters of non-binding recommendations spanning the full range of social, economic, and environmental issues (U.N., 1993). One chapter is devoted to the conservation of biological diversity, and biodiversity-related activities are featured throughout other chapters.

With agreements to conserve biodiversity, foster the sound utilization of forests, fisheries, agriculture, and other resources, transfer related technologies, share in financial investments, and the like, countries face the question: How can a nation determine what steps to take at home? Article 6 of the Convention calls for the parties to:

develop national strategies, plans or programmes, or adapt existing plans, to address the provisions of the convention; and to integrate biodiversity work into sectoral and cross-sectoral plans, programmes and policies.

The preparation of conservation and development strategies and related action plans is not new. Already, most countries have prepared a range of such exercises, including national conservation strategies, national environmental action plans, national development plans, and more recently, national sustainable development strategies, as well as sectoral plans for biological resources. Most already contain assessments of natural resources and have proposed strategic measures to strengthen administrative capability and improve conservation and use.

Experience with planning and implementing biodiversity-related measures has been limited in both scope and complexity. Indeed, most nations have already worked in national park planning, endangered species protection and recovery, and plant and animal propagation and breeding. Some countries have worked at larger scales to manage river basins and geographic regions, including biosphere reserves, and many have initiated rudimentary or advanced states of biotechnology. Yet, few countries have approached biodiversity planning and implementation in the comprehensive, integrated manner required by the Convention: from site and species protection, through seed and germ plasm collection, to international technological and financial cooperation, and biotechnology development.

These guidelines should help orient governments and NGOs, community and indigenous groups, and industry to how they might launch or expand biodiversity planning. The "illustrative biodiversity-planning process," offered here as a point of departure, has been drawn from the real-world experience of seventeen countries that have already taken on the challenge of Article 6. Each of these nations has pursued a unique path that reflects its particular cultural, political, ecological, and economic reality.

This study is targeted to help *biodiversity planners*—individuals who commit to collaborate with others to determine how to conserve their nation's biotic wealth, to use it sustainably, and to seek ways to share its values equitably. Such individuals will be agency personnel in forestry, agriculture, fisheries, national planning, foreign affairs, and finance; local government officials, community leaders, indigenous leaders, and NGOs; and representatives of resource management and extraction companies. These people are also *biodiversity implementors*—individuals committed to seeing that decisions made are properly acted upon, and that the proper conclusions are drawn from the experience.

Three types of biodiversity plans were called for in the Convention negotiations and the agreed articles: country studies, national strategies, and action plans. Details on the history of these important Convention decisions and the country study process are found in Annex B. All three are components of a larger and quite flexible process that can help countries build on existing institutions, programs, investments, and capabilities. This process is *cyclical.* It leads countries to periodically assess their biota and capacity, identify an evolving set of priorities and actions for responding to new opportunities, and prepare different reports to government, society, and the Convention on their findings and conclusions. It is *multi-sectoral,* involving all biodiversity-related government ministries, private resource-using industries, and civil groups that reside in, depend on, or, for other reasons, care deeply about the future of life on Earth. And finally, it is *adaptive.* It is revised and reformulated as new information arrives, and the results of previous activities and investments are continuously assessed.

Several complex and controversial aspects of biodiversity planning are being dealt with at this time through other efforts. These include establishing criteria and priorities for access to and use of the financial resources that are to be available to developing countries through the Convention funding mechanism, economic evaluation of biodiversity (UNEP, 1993; McNeely, 1988), and biodiversity and environmental impact assessment (Therivel et al., 1992; World Bank, 1991; Wilson, 1990). The Global Environment Facility (GEF) partners (The World Bank, United Nations Development Programme [UNDP] and UNEP) will soon release guidelines to help nations develop biodiversity projects. The results of ongoing deliberations are not second-guessed here. Prospective biodiversity planners are urged to obtain copies of the materials listed in Box 1, and to follow closely the continuing development of components of the Convention to obtain further guidance. Similarly, for background materials on the definitions, values, and nature of biodiversity and biological resources, readers will have to consult the various publications cited in the References section of this document.

Finally, there is no universal language or terminology in use for the various phases or steps in the biodiversity planning process. Various individual specialists and organizations attach different meanings to such words as *assessment, strategy,* and *action plan,* and to such process-oriented words as *planning, strategy development* and *implementation.* Here, the language is kept basic and simple, recognizing that (a) there is already a lexicon within the text of the Convention and (b) the native language of many readers and users of this document may be different from the language in which the report is presented. Thus, readers are urged to search for the contextual meaning of the words used and avoid bogging down in terminology issues.

BACKGROUND

The World Resources Institute (WRI), The World Conservation Union (IUCN) and the United Nations Environment Programme (UNEP) established a joint Biodiversity Program in 1989. Through an open process of technical workshops, regional dialogues, and research, the joint effort prepared and launched the *Global Biodiversity Strategy—Guidelines for Action to Save, Study, and Use Earth's Biotic Wealth Sustainably and Equitably* (WRI/IUCN/UNEP, 1992). The United Nations Food and Agriculture Organization (FAO), the United Nations Education, Scientific and Cultural Organization (UNESCO), and over 45 governmental agencies, NGOs, scientific institutes, and rural and indigenous communities contributed to formulation of the *Strategy*. Drafts of the *Strategy* were offered by the joint program to the inter-governmental negotiating process that led to the Convention on Biological Diversity. The *Strategy* has since served to inform and orient institutions and individuals worldwide and is now available in eight languages.

The Biodiversity Program continued its work with the preparation of a preliminary framework on *Biodiversity Indicators for Policy-Makers* (Reid, et al. 1993), *Biodiversity Prospecting: Using Genetic Resources for Sustainable Development* (Reid, et al. 1993b), and an array of regional workshops. The Global Biodiversity Forum series in 1993 complemented the Convention process. The Forum develops a broad-based public and private sector

BOX 1

Key Documentation that Should be Made available to Support Biodiversity Planning

Carew-Reid, Jeremy, Robert Prescott-Allen, Stephen Bass, and Barry-Dalal Clayton. 1994. *Strategies for National Sustainable Development—a handbook for their planning and implementation.* IUCN and IIED, Gland, Switzerland, and London.

Glowka, Lyle, Francoise Burhenne-Guilmin, Hugh Synge, Jeffrey McNeely, and Lothar Gundling. 1994. *A Guide to the Convention on Biological Diversity.* World Conservation Union (IUCN), Cambridge, U.K.

IUCN/UNEP/WWF. 1991 *Caring for the Earth: A Strategy for Sustainable Living.* Gland, Switzerland.

McNeely, J., K. Miller, W. Reid, R. Mittermeier, and T. Weiner. 1990. *Conserving the World's Biological Diversity.* IUCN, Gland, Switzerland; WRI, CI, WWF-US, and the World Bank, Washington, DC.

Therivel, Riki, Elizabeth Wilson, Stewart Thompson, Donna Heaney and David Pritchard. 1992. *Strategic Environmental Assessment.* Earthscan Publications Ltd., London, U.K.

United Nations. 1993. *Agenda 21, Rio Declaration, Forest Principles: Final Text of Agreements.* United Nations, New York.

United Nations Environment Programme. 1993. *Guidelines for Country Studies on Biological Diversity.* UNEP, Nairobi, Kenya.

United Nations Environment Programme. 1992. *Convention on Biological Diversity.* UNEP, Nairobi, Kenya.

World Conservation Monitoring Centre. 1992. *Global Biodiversity: Status of the Earth's living resources.* Chapman & Hall, London, UK.

WRI/IUCN/UNEP. 1992. *Global Biodiversity Strategy.* WRI, Washington DC.

Additional material should include national development plans, a completed country study if available, and national legislation regulating biological resources.

constituency of stakeholders related to biodiversity and biological resources and explores and debates critical issues.

Article 6 of the Convention, Resolution 2 of the Nairobi Final Act, and Chapter 15 of *Agenda 21*, all point to the central importance of the preparation of national biodiversity plans in an effort to assist countries in assessing the gaps in their conservation and development programs, in building the capacity to address strategically their biodiversity needs and opportunities, and in formulating their priorities for action. Starting in 1992, the joint Program began working closely with partner governments to observe and support initial planning efforts and to draw the lessons being learned from this pioneering experience. This publication presents the initial results of this cooperative research, analysis, and dialogue process.

METHOD OF THIS STUDY

This work began with a worldwide survey of recent and ongoing planning efforts that sought *inter alia* to address biodiversity goals. Such plans include national conservation strategies, national environment action plans, sustainable development plans, biodiversity country studies, strategies, and action plans.

WRI specifically looked for planning efforts that (a) are directed to help policy-makers and (b) are representative of the world's regions, including countries with economies in transition and small island states *(Annex C)*. Eighteen countries were invited by WRI to contribute 10-page case studies following a *pro forma* outline *(see Annex D)*, along with relevant maps and supporting documentation. In all, 17 countries provided such materials. Further, WRI and IUCN project staff visited and interviewed the lead organizations and individuals from these countries (with the exception of the South Pacific Region).

An illustrative approach for biodiversity planning was drawn from the analysis of submitted case study documents, country studies, national strategies and action plans, and personal interviews. The IUCN/IIED strategic planning handbook grounded the study in principles and procedures derived from extensive work by developing countries (*see Box 2*) (Carew-Reid et al., 1994). The obstacles noted in the country case studies were addressed throughout the analysis.

Reviews of the first unedited draft of the study in June 1994, along with the analysis and synthesis of the country cases, plans, and interviews, provided grist and insights for the second draft. At a four-day workshop in Williamsburg, Virginia, in September 1994, the second draft was rigorously peer reviewed. International participants at the workshop included those that contributed country case studies and others with experience in biodiversity planning. *(See Annex E.)* A third draft was prepared after the Williamsburg workshop for final review by UNEP, IUCN and WRI.

BOX 2

Strategies for Sustainable Development Handbook Series

These guidelines can be used in conjunction with the *Strategies for Sustainable Development Handbook Series* being produced by IUCN and its partners to assist countries and communities implement *Agenda 21*, and the action program of the United Nations Conference on Environment and Development (UNCED). The series will include handbooks on national strategies for sustainable development, local strategies, assessing progress toward sustainability, biodiversity action plans, indigenous peoples' involvement, integration of population and resource-use planning, and regular companion volumes of case studies addressing the key issues of concern to strategy implementation.

HOW TO USE THIS GUIDE

This guide can be used in various ways. As a reader, it can acquaint ministry staff, NGO and community leaders, and others with the basics of biodiversity planning. As a training tool in workshops and other capacity-building settings, it can help government and NGO representatives design a planning process.

Naturally, each country team will develop its own approach as it gains experience and identifies unique challenges. For this reason, this guide will have served its purpose when it is supplemented by new, more detailed, and creative materials and methods that match each country's political, cultural, economic, and ecological needs.

In the following chapter, the relationship among the three biodiversity-planning tools established by the Convention—country study, national strategy, and action plan—is explored along with associated negotiations and resolutions. These tools are then related to other planning efforts under way in most countries to illustrate opportunities for reducing overlap while incorporating biodiversity considerations into all relevant sectoral and national plans. In *Chapter 3,* lessons learned from the early experience of the seventeen countries, as reflected in case studies and interviews, are analyzed. *Chapter 4* presents a framework for national biodiversity planning using the Convention on Biological Diversity and identifies the specific actions needed. *Chapter 5* lays out an illustrative biodiversity-planning process based on seven fundamental steps. *Annex A* presents short profiles in national biodiversity planning that capture the experiences of the cooperating countries. *Annexes B-F* provide the reader with additional background information on the Convention, additional information on the methods of this study, details on the contributors to this report, and the text of selected articles of the Convention.

Chapter Two • Relationships among Biodiversity-Planning Tools and Between Strategic Biodiversity Planning and Other Planning Efforts

2

In this Chapter:

• *Introduction*

• *Relationships Among Biodiversity–Planning Tools*

• *Relationship Between Strategic Biodiversity Planning and Other Planning Efforts*

Relationships Among Biodiversity-Planning Tools and Between Strategic Biodiversity Planning and Other Planning Efforts

INTRODUCTION

In a growing number of countries, three basic types of biodiversity plans, or tools, are under preparation or further development—country studies, national strategies, and action plans. The divergent use of terminology can cause some confusion as countries communicate with each other. Experience suggests, however, that these three tools can be used in sequence as components of a larger planning methodology that also includes the integration of biodiversity considerations into the programs and budgets of major social and economic sectors.

RELATIONSHIPS AMONG BIODIVERSITY-PLANNING TOOLS

Three biodiversity-planning tools have already been established within the lexicon of the Convention and associated resolutions.

The country study

The country study is an assessment designed to gather information on the status and trends of the nation's species, genetic materials, and habitats and landscapes: on the status of current conservation and use mechanisms; and on the monetary and non-monetary costs and benefits involved, and to be a first overview of the opportunities and problems of protecting and mobilizing a country's biotic wealth.

The national strategy

The national strategy analyzes the descriptive data and information in the country study; identifies potential goals and objectives; and analyzes the gaps between current reality and the aspirations espoused in the objectives, the issues and opportunities for attaining the objectives, the environmental impacts of various options, and the implications for national human, institutional, infrastructural, and financial capacities and the possible need for international cooperation. The strategy proposes the action and investment needed to address each objective and assigns priorities to each.

The action plan

The action plan spells out the steps needed to implement the selected strategy, and then addresses practical questions: which organizations (public and private) will take up which activities, over what time frame, at what location, by what means, and with what resources? Specifically, which institution will be responsible for taking which protection and use measures, with what personnel, institutional resources, facilities, and funding? What will the implementation schedule be? And what, if any, international cooperation is needed and how will it be negotiated?

These tools spring from the negotiations that led up to the Convention and were further developed and re-enforced in the Convention itself. During negotiations, developed and developing countries were invited to assess their biodiversity and biological resources, to identify the costs and benefits associated with their management and conservation, and to estimate the resources needed to cover those actions and investments that the country may not be able to afford itself, given other established commitments. In most developing countries, these "country studies" were supported financially and technically by UNEP and the GEF or through bilateral assistance.

Countries that cooperated in this study felt that country studies (or "biodiversity assessments") provided several benefits. First, they enabled the country to involve diverse intellectual, informational, scientific, managerial, community, administrative, and industrial interests in a common effort that fostered awareness of the country's biotic wealth. Second, they set the stage for building the political will needed to sign and ratify the Convention and to implement it in the national self-interest. Third, they can be an invaluable tool for evaluating and monitoring action plans and biodiversity in general over the mid- to long-term. And, fourth, they provided important quantitative economic information that was used during the negotiations to evaluate the need for biodiversity investments around the world.

Article 6 of the Convention calls for parties to prepare national strategies, action plans, or programs, and to integrate biodiversity considerations into sectoral and other national plans. The experience of the cooperating countries demonstrates the usefulness of following-up the country study (assessment) with work that identifies strategically and comprehensively what needs to be done to meet stated objectives (strategy) and how that work can be accomplished (action plan). *(See Boxes 2 and 3.)* This study focuses mainly on the strategy and action plan in biodiversity planning.

The biodiversity plans of the 17 countries reviewed here vary in style and content, perhaps reflecting differences in culture, political reality, capacity, and biodiversity. They have not all followed a particular sequence—country study, national strategy, action plan—and in each country, the components mean somewhat different things. Still, as Box 4 and Chapter 5 make clear, there is much to be gained by employing the three tools in sequence.

The basic process outlined in Box 4 provides a flexible yet comprehensive approach to biodiversity planning. The three tools listed in sequence in the left-hand column relate to the seven basic planning steps shown in the second column. Clearly, current country studies correlate with the assessment step, current strategy studies with the analysis step, and action plans, to date, with the plan-of-action step. (Steps 1—getting started, with its vital preliminary decisions and organization, 6—monitoring to track impact and change, and 7—reporting to various constituencies may not be explicitly covered by the three tools.)

RELATIONSHIP BETWEEN STRATEGIC BIODIVERSITY PLANNING AND OTHER PLANNING EFFORTS

Those charged with biodiversity planning are already burdened with several planning commitments (MacKinnon, 1994). The individuals, agencies, and organizations involved are also involved in national environment plans, tropical forest action plans, national conservation strategies, protected areas systems plans, and individual site plans, among others. *(See Box 6.)* In addition, many nations have national development plans, national sectoral policies, and, more recently, national sustainable development plans or strategies created in response to *Agenda 21* and the Rio accords. These plans can be classified by scope (local, regional, national, or multi-country), sector (forestry, agriculture, fisheries, industry, education, etc.), or theme (water, desertification, population, biodiversity, etc.).

Biodiversity and biological resources are fundamental to food and nutrition, health, timber, energy, land use, and virtually all other areas of human welfare. Thus, as the framers of the Convention have made clear, though nations may find it useful to prepare separate biodiversity reports, they can find value in, and save steps by associating the biodiversity process with other national and sectoral planning efforts, thereby minimizing overlap and integrating biodiversity considerations into sector plans, programs, and budgets.

Two options could improve the efficiency and

BOX 3

The Biodiversity Planning Process: Expanding Participation and Benefits

Biodiversity planning is iterative and cyclical. Expanded participation of stakeholders allows them to share benefits and promote widespread commitment and action.

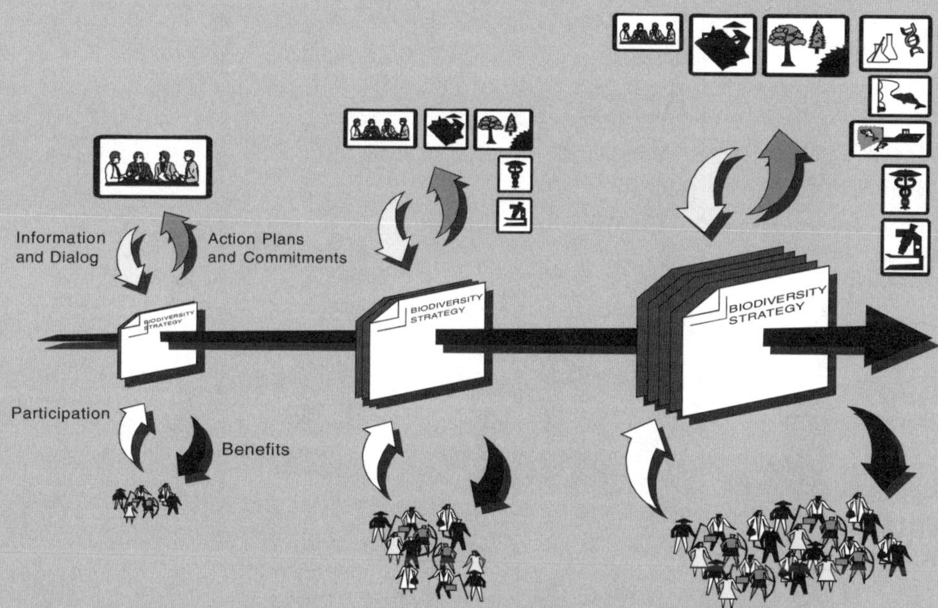

integration of biodiversity planning. One is to incorporate biodiversity planning into existing national plans, strategies or programs; and the other is to integrate biodiversity management into all sectors in a country that impact biodiversity.

INTEGRATION OF BIODIVERSITY PLANNING INTO EXISTING PLANS

The biodiversity planning team needs to determine whether existing plans, strategies and programs can be made more efficient through reducing overlap and eliminating redundancy. Most plans gather, analyze, and present a similar set of basic data and information about the geography of the country and its resources and peoples, the administrative workings of its institutions, its development goals and plans, and its financial situation. With all this data already available, why should a country commit itself to yet more planning? What distinguishes biodiversity planning from other types of plans, what does it add, and at what cost? This will help to determine how a country approaches its

BOX 4

Basic Steps for Biodiversity Planning and its Relationship to the Planning Tools: A Cyclical and Adaptive Process

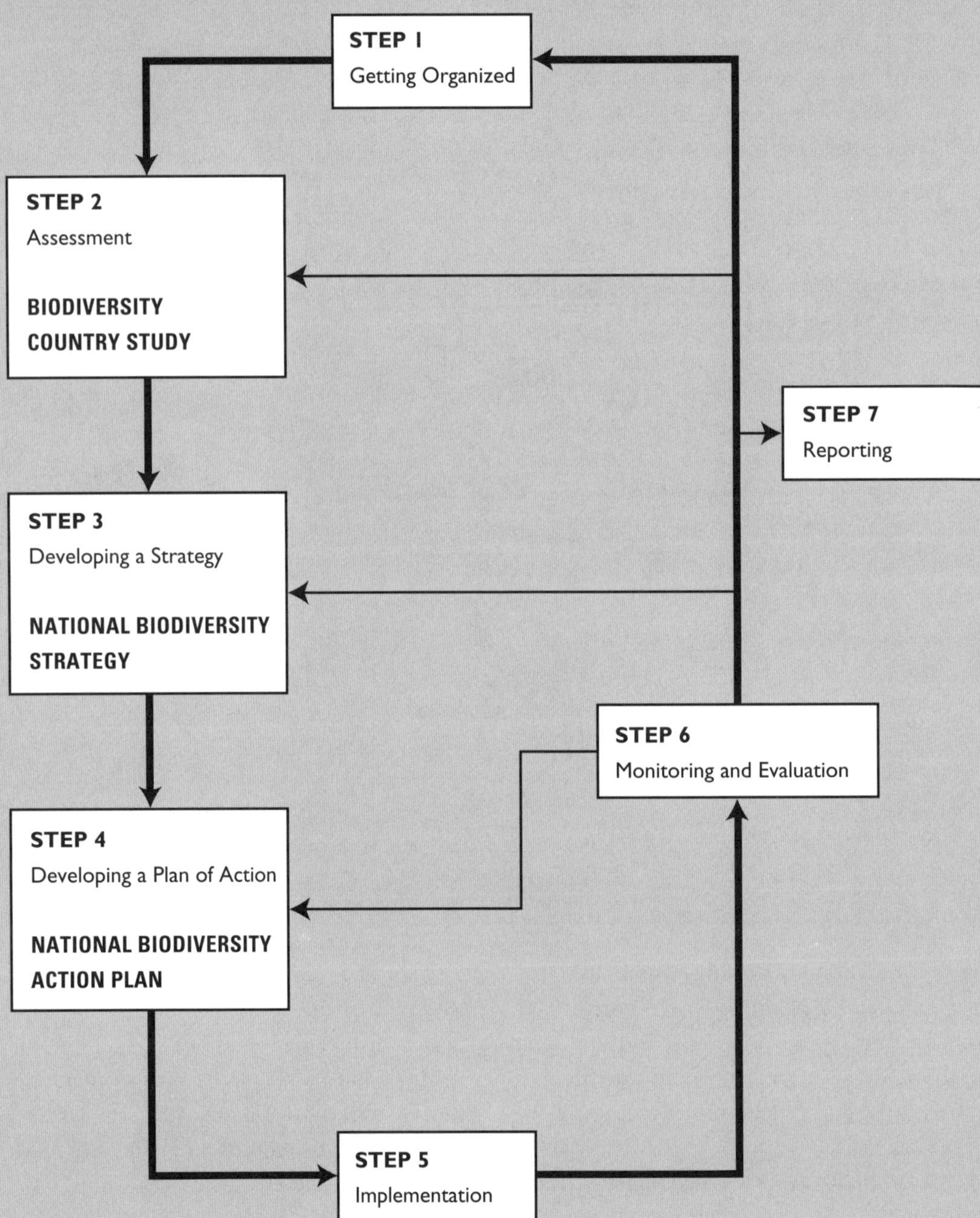

biodiversity planning.

A review of the Articles to the Convention illustrates the components to be addressed by a biodiversity strategy, action plan, or program. These components relate directly to the multiple-sector issues of health, nutrition, building materials, climate, water, the aesthetic and cultural well-being of a society, and to the equitable sharing of benefits from biological resources. Biodiversity concerns can be included in existing plans that currently address these issues.

Additional biodiversity components, such as measures for *in situ* conservation, inventory, capacity-building for biotechnology, and programs for technology cooperation, can be incorporated into the national development plan, a tropical forestry action plan, a national environmental action plan, or any other major planning tool. A separate biodiversity plan, *per se*, is not absolutely necessary to achieve the goals and meet the commitments of the Convention.

While experience to date in integrating bio-

BOX 5

Relationship Among Biodiversity Planning Tools, Planning Steps, and Outcomes

BIODIVERSITY PLANNING TOOLS	BIODIVERSITY PLANNING STEPS	OUTCOMES AND/OR RESULTS
	1. Getting organized	Political mandate, establishment of lead agency, funding, etc.
Country study	2. Assessment	Awareness; popular and political commitment; basic grasp of the task that lies ahead; initial data base; initial human and institutional capacity
National strategy	3. Developing a strategy	Goals and objectives; selected priority actions and investments
Action plan	4. Developing a plan of action	Decisions on who will do what, when, where and how, and with what human, institutional, facility and financial resources
	5. Implementation	Laying of groundwork; policy and legislative reform; environmental education; equitable distribution of benefits; technology cooperation, etc.
	6. Monitoring and evaluation	Knowledge of change in status and trends in biodiversity and biological resources.
	7. Reporting	Various reports including: • national executive, parliament and the people • country study (assessment) • national strategy • action plan or plans by sectors • periodic reviews of status and trends • reports to the Convention, Commission on Sustainable Development, etc.

diversity into existing plans is preliminary, it reveals that this type of integration is feasible. Emerging guidelines include:

1. ***Relate components to sectors***—place the various components of biodiversity into sectors that cover closely-related topics or issues. For example, habitat conservation may best be addressed as part of a land-use plan, a tropical forestry plan, etc.
2. ***Seek complementary linkages***—ensure that the linkages between biodiversity components and sector issues are complementary and mutually beneficial. For example, ensure that incentives for agricultural production do not foster land use practices that damage wildland habitats and key genetic resources.
3. ***Foster comprehensive action***—combine related components into comprehensive packages. For example, to develop the nation's biotechnology capability, ensure that actions and investments include technology cooperation, capacity building, and the appropriate policy and administrative reform to promote this goal.

The challenges to this approach include:

1. ***Commitment***—providing the biodiversity component with enough interest and commitment by the appropriate implementing agency and community to keep it from being buried in a broader program;
2. ***Awareness***—raising awareness of the importance and value of biodiversity where it might otherwise get lost among the array of other national and community priorities;
3. ***Connection to Convention Mechanisms***—achieving a solid budgetary commitment, including funding through the Convention's financial mechanism.

These challenges indicate that while it may be quite efficient to incorporate biodiversity components for action and investment into various other types of plans, the risk is that components could be lost in a field of other activities. This possibility does not, however, rule out the importance of incorporating biodiversity "thinking" into other plans.

INTEGRATION OF BIODIVERSITY PLANNING INTO MULTIPLE SECTORS

Current experience in Norway provides one approach for incorporating biodiversity considerations into the various sectors of government.

Each ministry influencing biodiversity in the Norwegian government was asked to prepare a sector-specific biodiversity strategy. Guidelines for sector biodiversity planning were provided to the ministries by the Norwegian Directorate for Nature Conservation. The guidelines suggest that each sector will face peculiar challenges and problems in dealing with biodiversity conservation and sustainable use. In each case, action will have to be based on a thorough analysis of existing information. *(See Box 6)*.

The profiles in Annex A show the very preliminary experience of 17 country partners in integrating biodiversity with other planning levels and topics. Five of these countries have made their biodiversity-planning activities components of larger development, environment, or sustainable development plans. Four studies were carried out under the mandates of a national sustainable development strategy: biodiversity was added as a component of sectoral assessments within a new integrated multi-sector development and environment plan. In the biodiversity-planning effort, biodiversity actions were integrated into existing sectoral plans and various national environmental strategies rather than into a separate biodiversity strategy.

The importance of multi-sectoral and multi-disciplinary approaches to biodiversity planning is widely emphasized here and in most other literature on biodiversity planning. The case study profiles in Annex A show how some countries have experimented with this new integrated dimension. However, because much of this methodology is in its infancy and its results are unproven, present guidelines are preliminary and general. As countries move ahead with this type of integration, more research is needed to provide accurate guidance.

Initially, biodiversity conservation may be better served by distinctive, highly focused strategies and action plans that promote awareness, unleash political will, and funding. Within a few years, however, biodiversity considerations may well be incorporated into normal governance, administration, resource use, and enterprise. Routine sector plans in forestry, agriculture, water resources, transportation, trade, and other areas may contribute to conservation, sustainability, and equity. Standard practice on farms, in forest and fishing operations, and in business may include biodiversity protection and use.

BOX 6

Linkages of Selected International, National, and Regional Environment and Development Strategies

(Adapted from Carew-Reid, et al. 1994)

GEOGRAPHIC SCOPE	SUBJECT SCOPE: MULTI-SECTORAL	SUBJECT SCOPE: SECTORAL OR THEMATIC
INTERNATIONAL	• Stockholm Conference Action Plan[2] • World Conservation Strategy[3] • Report of World Commission on Environment and Development (Our Common Future)[4] • Report of Latin America & Caribbean Commission on Development *(Our Own Agenda)*[5] • Caring for the Earth: A Strategy for Sustainable Living[6] • *Agenda 21*[7] • Strategies for Shared Regions (Regional Seas Programmes, River Basin Strategies, etc.)	• Global Biodiversity Strategy[8] • Tropical Forestry Action Programme[9] • Strategy and Action for Sustainable Agriculture and Rural Development[10] • Global Strategy for Health for All by the Year 2000[11] • Plan of Action to Combat Desertification[12] • World Population Action Plan[13] • International Environmental Education Programme[14] • Vancouver Action Plan for Human Settlements[15] • Mar del Plata Action Plan for Water Resources Development[16] • Strategy for the Protection of the Marine Environment[17] • Climate Change Strategy[18] • Global Plan of Action on Plant Genetic Resources for Food and Agriculture
NATIONAL[1]	• National Development Plans • National Conservation Strategies • National Environmental Action Plans • National Sustainable Development Strategies • Provincial Conservation and Sustainable Development Strategies	• Sectoral Master Plans • Tropical Forestry Action Plans • National Plans to Combat Desertification • National or provincial strategies and action plans on biodiversity, climate change, energy, environmental education, indigenous people, population, etc.
REGIONAL OR LOCAL	• Conservation/environmental/sustainable development strategies and action plans for political/administrative regions, natural regions, municipalities, etc.	• Regional or local strategies and action plans on biodiversity, climate change, energy, environmental education, indigenous peoples, population, etc.

[1] National includes provincial or equivalent strategies in countries with a federal system. [2] UN 1972; [3] IUCN/UNEP/WWF 1980; [4] WCED 1987; [5] UNDP/IADB 1990; [6] IUCN/UNEP/WWF 1991; [7] UNCED 1992; [8] WRI/IUCN/UNEP 1992; [9] FAO/WRI/IBRD 1987; [10] FAO 1991; [11] WHO 1981; [12] UNCOD 1977; [13] WPC 1974; [14] UNEP/UNESCO 1975; [15] UNCHS 1976; [16] UNWC 1977; [17] IMO 1983; [18] WMO/UNEP 1992.

BOX 7

Example of a Framework for Sectoral Plans

(Adapted from the Norwegian National Action Plan for Biological Diversity—Guidelines for Sectoral Plans [Directorate for Nature Conservation, 1994)

1. INTRODUCTION

2. IDENTIFICATION OF THE BASIS FOR ACTION

- identify the components of biodiversity (Article 7)
- identify the processes and categories of activities likely to significantly harm or block conservation and sustainable use of biodiversity (Article 7c)
- criteria for determining which components of diversity require conservation measures are (a) the degree to which they are threatened or vulnerable, and (2) their level of importance.

2.1 IDENTIFICATION OF PROCESSES AND ACTIVITIES

- review the current status of all processes and activities within each ministry's sphere that affects biodiversity, distinguishing between activities and processes initiated by the ministry itself, and those initiated by others but administered by the ministry. Where responsibilities for certain activities are shared or unclear, ministries should coordinate their work on sectoral plans.

Identify groups of activities that have an impact on biodiversity; provide neutral reviews of and quantify, as far as possible, physical, chemical and biological impacts on the environment and the motives underlying activities and their impacts.

- each ministry should base its review of ecosystems and habitats, species and communities, and genome and genes, to consider:
  - –the physical impacts on the environment, (whether from the modification of an area, extraction, or harvesting of biotic and abiotic resources)
  - –the chemical impacts on the environment (whether from the emissions of gases, or discharges of nutrients, organic compounds, heavy metals and waste and other materials)
  - –the biological impacts on the environment, (whether from the modification of organisms, the production of organisms, the accidental or intentional release of wild organisms, the domesticated and cultivated organisms, and gene-modified organisms)

2.2 IDENTIFICATION OF BIOLOGICAL DIVERSITY

More specifically,

- each ministry shall describe the biological diversity for which it is administratively responsible or which is affected by the processes and activities identified in item 2.1.

Identify wild and domesticated or cultivated biological diversity that is managed or affected at the ecosystem, species, intraspecies and genetic levels, and note the current and expected status of biodiversity in relation to the impact of factors identified previously.

- on this basis, the current and predicted status of biological diversity in relation to the effects of factors identified in 2.1 should be reviewed, including both negative and positive existing and potential impacts.
- each ministry should indicate the vulnerable or threatened components of biological diversity affected by the ministry's activities. The causes should be specified.
- each ministry should assess whether there are satisfactory mechanisms for evaluating the effects on biological diversity of the activities and processes for which the ministry is responsible.

2.3 IDENTIFICATION OF THE VALUE OF BIOLOGICAL DIVERSITY

The value of biodiversity is determined largely by interactions between human society and biodiversity. Its value, therefore, varies widely with economic, cultural, religious and ecological conditions, and also between the local, national, and regional levels.

- identify the type of value associated with the diversity recorded in accordance with item 2.2. Emphasise mainly the types of value that can be related directly or indirectly to the administrative responsibilities of each ministry.

Continued on page 19

Continued from page 18

- describe and, if possible, quantify the various types of value. Each ministry should assess whether there are satisfactory mechanisms for assessing the value of biological diversity related to the activities and processes identified according to item 2.1.

Identify, to the extent possible and relevant, the current and future value of biodiversity associated with life-sustaining ecological systems, use, recreation, science and education, aesthetic, symbolic, emotional and cultural aspects; and quantify to the extent possible and relevant, the actual and potential economic value associated with biological diversity.

2.4 IDENTIFICATION OF POLITICAL, LEGAL, ECONOMIC, AND ADMINISTRATIVE INSTRUMENTS

Convention Articles 8 through 14 describe a wide range of instruments and measures that the parties are to implement, as far as possible, to ensure the conservation and sustainable use of biodiversity. Each ministry is to review the range of instruments available:

Identify political, legal, economic and administrative instruments available for the conservation of biodiversity; mitigating or modifying impacts, processes, activities and underlying motives; the restoration of biological diversity; and improve the level of knowledge of biological diversity throughout society.

Each instrument should be related to a specific goal. The relevant instruments may be political, legal, economic, or administrative and include management of the ministries' subordinate agencies by means of directives, the distribution of management responsibilities, and the allocation of resources.

When considering economic instruments, such as those called for in Convention Article 11, for example, the review should include direct allocation of funds, subsidies, taxes and dues, as well as price policy, and it should include both incentive and dis-incentive measures.

3. SPECIFICATION OF GOALS

Each ministry must draw up an overall objective for its plan, based upon the results of the review in item 2 and define and discuss the obstacles to achieving this objective. Situations of uncertainty because of a lack of information and knowledge are to be considered within the precautionary principle from the Preamble to the Convention:

"...where there is a threat of significant reduction or loss of biological diversity, lack of full scientific certainty should not be used as a reason for postponing measures to avoid or minimize such a threat."

4. PROPOSALS FOR ACTION

4.1 STRATEGY

- each ministry is to draw up a strategy for measures to ensure the conservation of biodiversity on the basis of the information identified in item 2 and the goals and problems formulated in item 3.

4.2 ECONOMIC AND ADMINISTRATIVE CONSEQUENCES

- evaluate the economic and administrative consequences of the measures proposed in item 4. Include an assessment of the benefits of the measures proposed or the cost of not implementing them; this should reflect the full range of values.

4.3 PRIORITY

- the evaluation in 4.2 should be used to prioritize the measures proposed in item 4.1. This should be coordinated with other ministries and viewed in relation to overall national priorities.

5. IMPLEMENTATION AND ASSESSMENT OF RESULTS

Specify measures for implementing the sectoral plan, including a timetable, and division of labor and responsibilities. Specify how the results are to be assessed in relation to the goals and time limits set out in the plan (Convention Article 26).

Chapter Three • Lessons Learned from Early Experiences

In this Chapter:

- *Introduction*
- *Ten Guiding Principles*
- *Obstacles to Biodiversity Planning*
- *Ingredients for Success*
- *Other Lessons from the Country Cases*

Lessons Learned from Early Experiences

INTRODUCTION

However wide-ranging, the methods of biodiversity planning used by the case countries do provide the basis for an illustrative approach (methodology) that reflects their experience. Whatever the approach, each country has designated a public office or a nongovernmental organization (NGO) office to coordinate the inquiry into its biotic assets, its capacity, and its ongoing public and private programs in conservation and resource use. Each has also determined what gaps exist between current reality and goals. Each country has identified what needs to be done to close these gaps and has started or strengthened a program to set this action in motion.

Of course, countries have come to the biodiversity-planning task with diverse capabilities. As shown in Box 8, the standard breakdown of countries as developing, developed, transitional economies, and small island states provides a helpful framework for understanding certain generic strengths and weaknesses. Differences aside, some basic principles emerge from the case studies that, with proper adaptation, can probably help all countries.

Countries from all four categories encountered obstacles to biodiversity planning, but also identified the factors that were most helpful in facilitating their work. These obstacles and helpful factors are discussed below. A brief analysis of the case-country experience later in the chapter provides readers with an overall grasp of the key lessons learned.

TEN GUIDING PRINCIPLES

The World Conservation Union (IUCN) and the International Institute for Environment and Development (IIED) have pooled their decades of experience in working with developing countries to list ten lessons learned from strategy-planning activities. These principles, which are of direct relevance to biodiversity planning, are summarized in Box 9 (Carew-Reid et al., 1994).

To begin the process, biodiversity planners must envision their task within a larger framework that features two central dimensions. One is improving and maintaining human welfare and livelihoods; the other is using ecosystems in ways that are sustainable through conservation, restoration, and enhancement measures. At the same time, planners' efforts must point solidly to achieving tangible goals.

1. Biodiversity strategies, action plans, or programs should improve and maintain the well-being of people and the productivity and diversity of ecosystems.

The actions contained in the strategies, action plans or programs should seek to balance the needs

BOX 8

Characteristics that Distinguish the Capacity among Countries for Biodiversity Planning and Implementation

CATEGORY OF COUNTRY	CHARACTERISTICS OF IMPORTANCE FOR BIODIVERSITY PLANNING AND IMPLEMENTATION
DEVELOPING COUNTRIES	• Limited number of qualified personnel. • Limited experience in cross-sectoral work and multi-stakeholder planning. • Limited facilities for key tasks (e.g., inventory, data management, collections, research). • Extremely limited internal public financial resources and little tradition of private sector or individual contributions to conservation and development. • Information is scattered, or held beyond borders.
DEVELOPED COUNTRIES	• Highly developed institutions with complex structures; difficult to shift to new policies and practices and to reform jurisdictional overlap. • Highly qualified personnel, with authority to do specific activities, often with little flexibility to shift to other work. • Highly advanced facilities in science, education, *in situ* and *ex situ* conservation and restoration, etc.
COUNTRIES WITH ECONOMIES IN TRANSITION	• Highly qualified personnel who are underemployed in their professional fields. • Established institutions that are in transition, resulting in rapid shifts in resource management and landownership policies, unclear mandates, and limited authority for work on biodiversity. • Well-established facilities, including universities, laboratories, research stations. • Limited internal public or private financial resources.
SMALL ISLAND STATES	• Small number of qualified personnel in a limited set of fields (lack of management and research teams on key issues). • Traditional institutions that govern use and ownership of land often in conflict with modern pressures for use and access to resources. • Key ecosystem components, mainly marine and coastal, in many cases shared with other island states in the region, thereby requiring cooperative agreements on basic resources essential to livelihoods. • Development pressures rapidly deplete limited natural resource base.

and socioeconomic opportunities of people with the maintenance of healthy and diverse ecosystems.

2. Plans should contribute to the larger goal of sustainable development.

Biodiversity strategies should integrate socioeconomic and ecological perspectives into biodiversity-related policies, plans, and programs, and they should be integrated whenever possible with larger sustainable development initiatives.

3. Objectives should be selected from the full scope of the Biodiversity Convention.

Objectives should be few in number so that they can be achieved during a reasonable period, relate directly to the self-interest and common good of participants, and be comprehensive and coherent enough that efforts to realize them are not fragmented. They should be clearly defined and measurable so that progress can be assessed. Wish lists of projects intended to please everyone end up satisfying no one.

4. The biodiversity planning process must be adaptive and cyclical.

Biodiversity planning is a process, not an isolated event or product. The process is adaptive, developing as it goes along and responding to change. It is cyclical insofar as the main components are reiterated over several years. A strategy need not and should not try to do everything at once. It can grow in scope, ambition, and degree of participation as human, institutional, infrastructure, and financial capacities increase.

5. The process should be as participatory as possible.

Participation means sharing responsibility for the strategy and action plan and jointly undertaking the plan. Participants should be those who have the responsibility for, depend on, live within, or otherwise care for the variety of life and living resources. The participants will have the values, knowledge, technology, or institutions needed for successful implementation. By the same token, they will find their values, knowledge, technology, and institutions changing as a result of their involvement in the process.

Participants bring information to the strategy, ensuring that it is based on a common understanding of purpose, problems, and solutions. Participation is the most effective way of communicating the information on which the strategy is based, its objectives, and the actions to be taken. People who participate in the design and the decisions about what should be done are more likely to

BOX 9

Ten Guiding Principles for Biodiversity Planning

(adapted from Carew-Reid, et al. 1994)

1. Biodiversity strategies, action plans, or programs should improve and maintain the well-being of people and the productivity and diversity of ecosystems.
2. They should contribute to the larger goal of sustainable development.
3. Objectives should be selected from the full scope of the Biodiversity Convention.
4. The biodiversity-planning process must be adaptive and cyclical.
5. The process should be as participatory as possible.
6. Communication and negotiation must be the lifeblood of a biodiversity-planning process.
7. Biodiversity planning's success depends on decision making and action.
8. The biodiversity-planning process should be integrated into each country's decision-making system.
9. The capacity for biodiversity-planning needs to be built at the earliest stage of the process.
10. External agencies should be "on tap," not on top.

understand the purpose of the actions and to implement them fully.

Participation should be expanded as the biodiversity-planning process develops. Initially, the strategy team may begin with selected representatives of government agencies, communities, business and industry, and indigenous groups. Gradually, previously unknown or unempowered but highly qualified and motivated individuals and groups will emerge to take part.

6. Communication and negotiation must be the lifeblood of a biodiversity-planning process.

Through communication and negotiation, participants explore their values, perceptions, and interests regarding their ecosystems, resources, economy, and society and debate and agree upon goals, objectives, and actions. Values are changed or strengthened, and knowledge is imparted. Participants inform others about the strategy. Accordingly, communication and negotiation methods need to be planned carefully as an integral part of the process.

7. Biodiversity planning's success depends on decision making and action.

Decision making is a central activity of the planning process, but the result is much more than a "plan" or set of reports. Biodiversity planning is a process of developing a long-term vision and sense of direction; carefully designing and selecting the key steps that can be taken in that direction; engaging everyone concerned—citizens' groups, communities, business, and governments—to carry them out; following through to ensure that the activities, investments, and changes called for by the process are actually realized; and, finally, extracting and disseminating the lessons learned.

The main components of biodiversity planning are assessment of issues and options; analysis and selection of goals, objectives, and alternative policies and actions; implementation of the actions called for; and monitoring and evaluation to keep strategy implementation on course and to enable it to adapt to changing conditions and results.

Clearly, the planning process goes well beyond the preparation of an assessment, an analysis of alternative policies and actions, or a plan of action. Rather, it is a means of making decisions and taking actions to strengthen or change values, perceptions, knowledge, technologies, and institutions. By the same token, a document or report is an essential tool for communicating the agreed-on goals, objectives, actions, and policies, as well as the mechanisms for implementation, monitoring, and reporting. But the plan or report is only a tool; it is not the strategy or action plan. If too much energy is spent preparing a document, too little may be left to promote or implement it.

8. The biodiversity-planning process should be integrated into each country's decision-making system.

The process should be integrated with other conventional development cycles, not just as an "add on" or a new chapter to old plans.

Biodiversity planning should build on priority concerns to which governments and people are already committed. Politicians and communities need to see the benefits and relevance of the process. It should draw on local knowledge, values, skills, and intuitions.

A biodiversity program should also build on past or current plans, rather than ignore or replace them, recognizing and capturing the best of what is available and has already been done.

9. The capacity for biodiversity planning needs to be built at the earliest stage of the process.

At a national level, this means building the capacity for cross-sectoral action, finding ways to integrate environmental concerns with development, and developing processes to alert government agencies and the private sector to their environmental responsibilities.

10. External agencies should be "on tap," not on top.

External financial and technical assistance should help the country concerned to increase its own capacity to undertake biodiversity planning. Governments that receive assistance must be able to take the lead in coordinating it. Locally designed and driven approaches to biodiversity planning should be favored over conditions on aid or notions of "model" or *pro forma* plans. Low-level continuous support over many years is almost always better than high-level support for a short time. Donors should support capacity-building, not just formulate plans. A key here is refocusing current investments, as well as making new ones.

Even if the planner applies all these principles, obstacles can be expected to arise that divert time, funding, and energy, and thus threaten progress. A look at the obstacles identified by the 17 partner countries who submitted case studies provides insights into issues to be examined and pitfalls to be avoided.

OBSTACLES TO BIODIVERSITY PLANNING

The most commonly mentioned obstacle to the preparation of country studies, strategies, and action plans is conflict over establishing the lead agency for a country planning effort. Too often agencies and ministries get into "turf battles" over who has the mandate to oversee the planning activities, and they fail to cooperate. Apparently, such territorialism is often motivated by competition for control of the anticipated financial flows of money from donors or central government.

Institutional obstacles noted include:

- difficulty coordinating and integrating numerous stakeholders and their respective issues
- poor coordination among government agencies and NGOs
- lack of provincial and local perspectives in planning
- lack of private sector involvement
- difficulty of building interagency consensus
- lack of trained scientific and managerial personnel
- poor collaboration between the donor project team and the in-country team
- lack of awareness of government agencies and local people
- expense of ensuring broad-based, multistakeholder participation
- lack of communication between the scientific community and policy-makers
- continual institutional change with economic restructuring

Scientific obstacles include:

- lack of research on biodiversity's role in ecosystems
- lack of sufficient scientific and economic data
- lack of trained biosystematists
- lack of information-management capacity
- duplication of scientific efforts

Obstacles mentioned in legal and policy areas include:

- lack of a strong policy framework and political commitment to its implementation
- lack of data to support policy work
- lack of capacity for policy analysis
- lack of integration of environment and development in national planning
- lack of well-formulated environmental laws and regulations
- need for economists' input
- difficulty in determining the costs of biodiversity conservation
- lack of clear policies on land tenure
- difficulty integrating indigenous land claims and interests into planning

The country cases point to activities designed to address the obstacles that have been encountered. These measures provide guidelines that may be valuable to other biodiversity planners.

The case studies provided the following actions or recommendations:

- Establish fellowships for overseas training programs to develop the nation's scientific and managerial capacity.
- Establish a steering committee to identify experts in other ministries and agencies or hire local or expatriate consultants to mobilize current government departments.
- Include representatives of indigenous peoples on steering committees.
- Establish a nonpartisan scientific committee to develop the first-draft strategy or action plan as a way to get around interagency disagreements and turf battles.
- Establish a special working group to facilitate negotiations with indigenous and non-indigenous landowners, especially for habitats and ecosystems of high national interest.
- Host regional and local workshops to build awareness and capacity at these levels; prepare provincial or state-level strategies to promote awareness, capacity, and participation.
- To promote greater integration of regional and local perspectives and to foster the development of jurisdictional positions on critical issues, establish interdepartmental government mechanisms in each province for reviewing the national strategy or action plan.

- Establish an advisory group to compile and integrate various stakeholder concerns and issues.
- Use conference calls, fax, and electronic communication to reduce the high cost of a broad participatory process.
- Develop public awareness campaigns to increase public and government understanding of the concept of biodiversity.
- Support research on mechanisms and approaches to conserve biodiversity (such as rapid biodiversity assessments and bioregional case studies).
- Support multi-divisional or multi-sector research programs to expand opportunities for collaboration.

INGREDIENTS FOR SUCCESS

Various factors identified by the case countries served to facilitate their work on country studies, strategies, and action plans:

- strong leadership by the lead agency
- support from participating government agencies
- strong local NGOs, communities, academic institutions, and civil organizations to provide broad and wide-ranging perspectives
- high international profile provided by the lead-up process to UNCED—a catalyst to in-country biodiversity activities
- signature and ratification of the Convention
- "critical impoverished" status of biodiversity, which can rally action
- a strong interdepartmental, interagency, interconstituent steering committee to provide a coherent driving force
- earlier plans and initiatives (such as country studies, national conservation strategies, forest sector plans)
- good ecological data and land-use maps
- available funds
- coherent and clear environmental laws establishing the mandate for developing a national strategy or action plan
- experience in multi-stakeholder policy negotiations
- strong regional government support
- involvement of people responsible for implementing any plans
- a national NGO forum or working group to support the strategy
- qualified personnel in government agencies
- a supportive high-level political environment
- broad community participation and a sense of ownership

Along with the obstacles encountered, the actions needed, and the factors that make corrective action possible, the case studies reveal a variety of other interesing lessons.

OTHER LESSONS FROM THE COUNTRY CASES

Following is a summary and analysis of responses from all the countries that submitted case studies on their efforts to develop a national biodiversity strategy, action plan, country study, or biodiversity program.

I. Institutional Base, Mandate, Level of Authority, and Opportunities for Endorsement

Of the countries that submitted case studies on national biodiversity planning, six are producing what they call *national biodiversity action plans*—the focus of their case studies. Five now have or are producing *national biodiversity strategies*—also the focus of their case studies. Five countries' case studies focus on their preparation of UNEP-sponsored *country studies on biodiversity*. One case study covers a regional *program* of cooperation among small island states to develop a comprehensive system of protected areas in the South Pacific.

In most case study profiles, the strategies and action plans cannot be substantively distinguished. They often have the same purposes, goals, and contents. In other arenas, efforts dubbed strategies generally propose a concrete plan for action, and efforts dubbed action plans usually cover the strategic planning that determines the countries' goals and operational objectives, as well as conservation gap analysis. In the methodology proposed in this guide, creating a strategy and an action plan are two distinct and sequential phases of an overall exercise in national biodiversity planning—a distinction that a few countries currently make. *(See Annex A)*.

In all cases of national biodiversity planning except one, the mandate and authority for initiating the country study, strategy, action plan or program was held by the central or federal government.

Ministry-level agencies were the direct lead agency in 12 countries, subministry-level government agencies played the lead role in three countries. One country study was produced by an NGO under contract from a ministry-level agency, and one program was developed by a United Nations (UN) agency and adopted by several coordinating-government ministry-level agencies.

All planning efforts in the case studies had the mandate and possibility of endorsement by each country's highest decisionmaking bodies as an "official" strategy, action plan, or program.

2. *Methodology of Country Study, Strategy, Action Plan, or Program*

The process and method for preparing and producing a country study, strategy, action plan, or program were similar in all 17 countries. For all three types of effort, the basic methodology consisted of the following steps:

- establishment of a multi-sectoral, multi-stakeholder steering committee to frame the issues and develop an outline structure
- creation of one or more working groups for specific issues
- development of a first draft
- staging of one or more workshops open to a broad range of stakeholders to discuss the draft
- revision of the draft based on the results of workshop consultations
- completion of the draft and efforts toward official recognition.

As expected, this basic methodology varied from country to country. All countries created a steering committee except for The Netherlands during its production of a biodiversity action plan, and Norway during its preparation of a strategy. What varied most was the number of consultations, working groups, and workshops. Generally, the steering committee and working groups produced a first draft before the document was released for wide public discussion. In the case of the South Pacific Biodiversity Conservation Programme, the program structure was developed by the UN South Pacific Regional Environment Programme and adopted by the national governments of the region.

The most notable variation on the above methodology was Norway's strategy process. Its methodology requires all federal ministries that have an impact on biodiversity to prepare sector-specific strategies that the Ministry of Environment must later integrate into one coherent national biodiversity strategy. This novel approach is intended to foster a greater sense of responsibility and ownership of a final strategy by the various ministries.

All 17 efforts reflected a high level of both multisectoral and intragovernmental involvement, as well as significant nongovernmental participation. NGO participation in various levels of production was high and seemingly limited only by capacity in three countries. Many countries noted that the openness for public debate and input was significantly greater than in previous efforts with similar policy exercises.

Significant differences emerged, however, in the relationship between the UNEP-sponsored country study and the development of a national strategy or action plan. Twelve countries specifically mentioned the positive impact of UNEP country studies on their current planning efforts. Seven countries completed countries studies *before* developing a strategy or action plan and used the Study as background in the later preparation of a strategy or action plan. The country studies provided data to use in setting and supporting objectives and actions, as well as a framework for the overall issues of the strategy or action plan. The methodology for all of the UNEP country studies was structured primarily using guidelines supplied by UNEP.

Four countries have prepared, or are preparing, a UNEP country study *after* the completion or near completion of a national strategy or action plan. Costa Rica had completed a national sustainable development strategy that included a biodiversity strategy by the time it was asked to produce a country study. For various reasons, other countries also prepared country studies *after* strategies or action plans were developed. Several countries were simply more advanced in developing national biodiversity-planning strategies than other countries and had commenced similar exercises before the Convention process dominated the biodiversity-planning agenda. In these cases, the country studies had little or no influence on the national strategy or action plan process or content.

The United Kingdom prepared a biodiversity action plan that reportedly contains both a country study and a strategy. All three exercises were included in one document because the country has ample research on biodiversity. Clearly, various nations interpret the scope and purpose of the three planning tools in various ways and adapt them as they see fit for their specific purposes.

Sources for guiding the methodology of strategies, action plans, and programs included the *Convention on Biodiversity,* the *Global Biodiversity Strategy,* previous *national conservation strategies,* previous country environmental legislation, and, for some, the experience of a foreign consultant.

3. Participation in the Development of the Strategy, Action Plan, Country Study, or Program

Common to all planning efforts examined was a strong commitment to a broad-based, multi-sectoral, decentralized, and multi-stakeholder participatory process in nearly all stages of planning. That said, the actual level of participation varied by country, though most of the countries' governments intended to include a high level of outside inputs.

All case studies except one indicated that the participation of federal ministries, subministries, parastatal organizations (i.e., relatively independent agencies with national responsibilities that are officially within a government's purview), universities, research institutes, the private sector, the general public, indigenous communities, and NGOs was sought. Broad participation was allowed through various mechanisms at various stages. In the preparatory stages, all efforts involving a steering committee included government officials, academics, scientists, and NGO representatives. The degree to which government dominated these committees varied from very high to seemingly low. To review early drafts, virtually all countries held consultative meetings or workshops - regionally, nationally, or both - or sent letters to stakeholders soliciting comments. One country study on a tight deadline followed a narrower process: it was prepared mainly by a committee of scientists and academicians.

Other mechanisms for broad participation included scientific panels, national meetings, and international conferences. The Netherlands conducted a two-day workshop for Dutch Embassy officials to discuss possible Dutch contributions to conserve biodiversity abroad. The South Pacific Biodiversity Conservation Programme sought the input of indigenous communities through the agency responsible for indigenous affairs.

Generally, NGO participation was strongest at all levels of planning in countries where NGO activity was advanced and vocal—particularly, Australia, Canada, Indonesia, Norway, and the Philippines. Some countries in which NGO participation was welcome did not have an active and effective NGO community. On the other hand, in Poland, an NGO was contracted by the government to prepare the country study because the government lacked the needed capacity; the views of the general public were not sought and other NGOs showed little interest in participating. Several countries tried to involve the private business sector but had trouble interesting this community. Indeed, only one country mentioned high private sector participation.

The actual influence of the many broad-based, multi-sectoral groups, and stakeholders on the development of the planning process and the final product cannot be determined from the case studies. As mentioned above, although every country stated that it was committed to and encouraged a broad participatory process, some countries no doubt were more successful than others in incorporating a broad array of views.

4. Goals and Objectives

The general goals and objectives were very similar for the six national biodiversity action plans, the five national biodiversity strategies, and the five country studies analyzed in the case studies. Every country couched its goals and objectives in terms of protection, sustainable use and equitable sharing of benefits—the stated goals and objectives of the Convention on Biodiversity—with several also referring to knowledge and information. Those shared goals were:

- catalyze immediate action to slow the rate of biodiversity loss and to develop a strategy that allows the sustainable utilization of natural resources while conserving biodiversity and the natural resource base
- conserve and enhance biodiversity within the country and contribute to global biodiversity conservation through all appropriate mechanisms
- guide domestic biodiversity conservation endeavors down to the provincial-level action

- help international agencies understand what the priorities for investment are
- ensure the sustainable use of nature's reserves.

Specific approaches for achieving these goals and operational objectives include:

- reinforce *in situ* conservation in nature reserves, parks, and protected areas and reinforce such conservation outside these areas as well
- increase public awareness of and involvement in conserving biodiversity
- improve the knowledge base about biodiversity.

The following goals and objectives of the country studies were taken from the UNEP guidelines for country-study preparation:

- identifying the current state of knowledge about biodiversity
- identifying important gaps in knowledge and assessing further needs and their costs
- identifying current pressures on biodiversity and future trends
- assessing the present and future value to humanity of biodiversity in the country
- assessing the cost of conserving biodiversity in the country
- identifying conservation priorities
- identifying appropriate implementation methods.

The goal for the South Pacific Biodiversity Conservation Programme is to protect the region's biodiversity by establishing conservation areas.

5. Relationship to National Development Planning

Because few strategies and action plans are completed and many more are in progress, it is too early to analyze their effect on central, long-term development planning. However, the case studies have illuminated various degrees of integration of the strategies, action plans, and country studies into larger, multi-sector planning. Some biodiversity-planning exercises intended to support *directly,* or formally, a larger, multi-sector national development or environment plan or strategy, whereas others were intended to have an *indirect,* or informal, influence on national development planning.

Five countries' biodiversity-planning activities outlined in the case studies were specific components of a larger development, environment, or sustainable development plan. These directly-linked efforts included two national biodiversity action plans, two national biodiversity strategies, and one country study on biodiversity.

Four of the planning efforts were mandated by, or closely associated with, a national sustainable development strategy. These biodiversity-planning efforts were one component of an integrated, multi-sector development and environment plan. The Australian strategy process was a distinct exercise, but planners referred extensively to the national sustainable development strategy. One biodiversity action plan was directly linked to larger planning efforts and sought to integrate biodiversity actions into several current national environmental strategies instead of developing a separate biodiversity strategy.

For strategies, action plans, country studies, and programs not directly linked to a larger national development plan, the degree of influence on national development planning varied. Decision-makers in several countries, seeking guidance on the focus, content, and language of national planning documents and legislation, have drawn on strategies, action plans, and country studies. Many agencies are also charged with improving the integration of existing development and environmental plans with biodiversity planning. These agencies could prove highly influential in planning.

For one country with a formerly centrally-planned economy, the immediate opportunity for influencing national planning for biodiversity is low. The transition to democracy has ironically hindered the government's ability to carry out strategic environmental planning, and the country lacks the mechanisms needed to support the development of a long-term biodiversity strategy. Coordination between development and environmental planning is also currently very poor.

6. Intended Target of Planning Efforts

All strategies, action plans, country studies, and programs were aimed at decision- and policy-makers. The case studies clearly indicated that the purpose of planning was to provide local, regional, and federal decision-makers with policy guidance on management options and priorities for investment and action. Some strategies tried to broaden the range of targeted decision-makers to include heads of nongovernmental organizations, and industry and community leaders. Some aimed to give donors an investment agenda. One established a task force to encourage and focus action outside of government. To meet decision-makers' needs, most efforts included policy-makers in most stages of the preparation of the strategy, action plan, or program.

7. Relationship of Strategies, Action Plans, Country Studies, and Programs to the Convention on Biodiversity

How fully the Convention's general provisions and specific articles are integrated into national biodiversity strategies, action plans, country studies, and programs depends on when the country planning efforts took place. Although countries whose biodiversity-planning activities *preceded* the Convention negotiations could obviously make no specific reference to the articles of the Convention, most such planning efforts *during or after* the final Convention negotiations were structured around its provisions or specific articles.

In seven countries, the strategies, action plans, country studies, or programs were *not directly* correlated to the articles of the Convention. Since the plans were formulated and developed *before* Convention negotiations were completed, specific articles could not be addressed. Yet, the active participation of country delegates in the Convention negotiations; the familiarity of country planners with the Convention's nature, content, and general objectives; and access to UNEP-supplied guidelines for preparing the strategies, action plans, country studies, or programs meant that Convention objectives were very effectively incorporated into plans. (The fact that the seven early planning exercises commenced *before* the Convention was adopted reveals a high level of commitment and planning sophistication.) Some of the 17 countries discussed here revised their earlier planning efforts to include and accommodate the articles of the Convention.

Two country study efforts based on the UNEP guidelines *indirectly* also addressed the contents and objectives of the Convention *before* the Convention text was finalized.

Ten countries structured their strategies, action plans, and country studies around the specific provisions or articles of the Convention or referred explicitly to the Convention. In one case, a draft strategy or action plan completed before the articles were adopted was modified to specifically accommodate the articles following final adoption of the Convention text.

For countries that started their planning process after the Convention had been adopted and signed in Rio, planners used the Convention provisions, structure, and articles as the basis for their strategy and action plan. Most of these countries structured chapters around specific articles or groups of related articles, proposing specific actions on each. The structure and language of the agreed text of the Convention made it easy both to model the strategy or action plan after the Convention and to address the specific articles.

The case studies have vividly indicated that directly linking strategies, action plans, and programs to the Convention has been quite successful. All planners recognized the Convention as a central focal point for structuring action at home on biodiversity. All countries indicated that future planning and revisions of biodiversity plans will reflect and incorporate the objectives and goals of the Convention. Further, because nearly all countries structured their strategies, action plans, country studies, and programs around the articles of the Convention, revised current plans to reflect the objectives of the Convention, or had previously incorporated Convention objectives into their national planning, most national action and investment in biodiversity will more than likely reflect the goals of the Convention.

However, two suggestions for improving the linkage between strategies, action plans, and country studies and the Convention were proposed. First, because the Convention's structure lends itself to the structure of strategies and action plans, national reports to the Convention will be facilitated. With simple modifications, the strategies and action plans could be used as the national reporting mechanism to the Convention. Second, the biodiversity issues and measures identified in Chapter 15 and throughout *Agenda 21* should be incorporated into strategies and action plans.

8. Scales of Planning Efforts

Not surprisingly, the main focus of all the strategies, action plans, and country studies was to establish *national* objectives and goals to orient activities for biodiversity management. Within a national framework, however, nearly all the strategies, action plans, and country studies focused on and encouraged local and regional planning and action. As an experimental methodology, Norway is developing several *local* biodiversity action plans to investigate possible mechanisms for implementing the national biodiversity strategy.

Although many countries specifically addressed their national responsibilities under other international environmental agreements—such as CITES, the Ramsar Convention, and the World Heritage Convention—several countries' strategies, action plans, country studies, or programs identify specific opportunities for *trans-boundary cooperation* for biodiversity management. A few countries analyzed their ongoing financial and technical assistance to biodiversity management in other countries.

No countries profiled in this study explicitly described their efforts to collaborate with neighboring countries in developing their strategy, action plan, or country study. However, Kenya, Tanzania, and Uganda were taking initial steps to share information, share technology, and prepare joint biodiversity plans.

It is apparent from several case studies that planning efforts addressed biodiversity management —in some capacity—at three ecological scales: *ecosystems, species,* and *genes.* In addition, because almost all planning efforts addressed the goals and objectives of the Convention—either directly or indirectly—most will probably address measures for the conservation of ecosystems, species, and genes.

9. International Assistance

Financial and technical assistance to countries for preparing national biodiversity strategies, action plans, country studies, and programs is common. The case studies show that financial support, usually including varying degrees of technical assistance, was provided to developing countries through various mechanisms.

Ten of the 17 countries profiled noted receiving financial assistance for direct or indirect support for developing national biodiversity plans. The World Bank was the most prevalent source of support for planning. Through technical-assistance grants and the Global Environment Facility (GEF), it supported three national biodiversity action plans, one national biodiversity strategy, one national environmental action plan, and one regional biodiversity program. World Bank financial assistance to these planning activities was in most instances one component of a larger financial and technical support program to a ministry of environment or natural resources. Seven countries received no foreign financial assistance. In most of the supported projects, the World Bank also contributed significant technical assistance for the strategy, action plan, or program.

Key bilateral donors were the governments of Australia, Canada, Norway, the United Kingdom, and the United States.

For the development of country studies, the United Nations Environment Programme (UNEP) provided financial and technical assistance, mainly in the form of written guidelines for study preparation, along with input by UNEP personnel.

Many international nongovernment organizations and research institutes also provided technical assistance in various planning efforts: the Conservation Foundation (CF), Conservation International (CI), the London Environmental Economic Centre (LEEC), the Nature Conservancy (TNC), the World Conservation Monitoring Centre (WCMC), the World Conservation Union (IUCN), World Resources Institute (WRI), and World Wildlife Fund-US (WWF).

Several countries noted that though much international financial and technical assistance supported national planning, the processes were not "donor driven." Rather, most countries followed their own instincts about process and content, and local policy experts and scientists most often played the leading role in developing the strategy, action plan, country study, or program.

10. Monitoring and Follow-up

All countries noted the importance of monitoring and follow-up for their strategy, action plan, country study, or program. All also acknowledged the need to formulate actions for implementation, identify monitoring components, and assign responsibility for implementation. Still, some countries spelled out actions, responsibilities, and monitoring needs in greater detail than others. Eleven included specific mechanisms for monitoring and follow-up in planning documents:

- the creation of a biodiversity data information center
- the establishment of a monitoring committee
- five-year reviews integrated into current government reporting mechanisms
- sector-specific monitoring programs
- public awareness campaigns
- the establishment of a national network of data collectors
- the publication of a comprehensive report on wildlife status every four to five years
- the preparation of sub-strategies for implementing specific strategy components.

Six country planning efforts have not progressed to the stage of establishing monitoring and follow-up policies or did not refer to this component. Responsibility for monitoring and follow-up, or for determining responsibilities, most often lies within the ministry charged with preparing the strategy, action plan, country study, or program. Where multi-ministry committees take the lead, they are responsible for developing policies for monitoring and follow-up and for assigning tasks and responsibilities to other ministries, agencies, and organizations.

11. Estimation of Cost, Personnel, and Time Required for Planning Efforts

Many countries provided some basic data on the cost, personnel, and time requirements for preparing their strategies, action plans, country studies, and programs. Most such estimates are very rough and may not include all costs. Some countries provided data only on the number of full-time staff employed for the project; others provided data on all people consulted. Reported expenditures on the planning processes range from US$10,000 for preparing an action plan in Chile to C$1.4 million for preparing a strategy in Canada; from requiring four people to prepare an action plan in The Netherlands to 500 people to prepare a biodiversity-conservation program in the South Pacific; from taking three months to prepare a country study in Poland to 42 months to put together a strategy in Australia.

Chapter Four • A Framework for National Biodiversity Planning: The Convention on Biological Diversity

In this Chapter:

- *Introduction*
- *Objectives of the Convention*
- *Checklist of Action Points*
- *Provisions of the Convention, in Brief*

A Framework for National Biodiversity Planning: The Convention on Biological Diversity

INTRODUCTION

The Convention on Biological Diversity (hereafter, the Biodiversity Convention or the Convention), opened for signature at the Earth Summit in Rio de Janeiro in 1992, can be used to frame the biodiversity planning process. The Convention articles provide the basic objectives for the process and the specific types of actions that countries can consider adapting to their needs, laws, and policies.

The plans prepared by the countries cooperating in this study *(see References)* address the objectives and action points covered by the Convention. Most plans prepared after the Rio Summit frame their planning process according to the substantive articles of the Convention. But those plans that pre-date Rio cover similar topics and issues, and many were revised after the Summit to bring both structure and content into line with the Convention articles.

The objectives of the Convention and a framework for action based on the international agreement follow. To focus on operational points, some liberty has been taken here to condense the official text. For the same reason, biodiversity planners may find it useful to prepare abbreviated guidelines drawn from the Convention articles. The sample checklist of action points presented below identifies the key operational elements to be addressed by the process. The World Conservation Union (IUCN) *Guide to the Convention on Biological Diversity* (Glowka et al., 1994) is also useful for this purpose.

OBJECTIVES OF THE CONVENTION

The objectives of the Convention as stated in Article 1 are:

- *conservation of biological diversity*
- *sustainable use of its components*
- *fair and equitable sharing of the benefits arising out of the utilization of genetic resources (through, among other channels, appropriate access to genetic resources and appropriate technology cooperation, taking into account all rights over those resources and technologies, and by appropriate funding).*

CHECKLIST OF ACTION POINTS

Biodiversity planners can draw from the Convention a basic set of issues to consider during the process:

- principles relevant to the particular country, including its culture, government, economic history, and distribution of biotic wealth
- assessment of the status and trends of biodiversity and biological resources and of the country's institutional, human, facility, and financial capacity

- mechanisms for *in situ* conservation and restoration
- mechanisms for *ex situ* conservation
- mechanisms for promoting the sustainable use of forest, range, agricultural, wetland, mountain, coastal, marine, and other ecosystems for the full range of their material and nonmaterial values—around protected areas and across the wider landscapes or waterscapes, especially at bioregional scales
- promotion of equitable sharing of the costs and benefits arising from the conservation of biodiversity and the use of biological resources
- measures to provide incentives to reduce consumption and waste
- inventory, collections, and systematic data management on species, genetic resources, habitats, and ecosystems
- research to better understand the role and function of biodiversity, ecological processes, and the sustainable use of biological resources
- measures to develop human, institutional, infrastructural, and financial capacity
- public awareness and education to develop popular and political will and commitment
- assessment of the impacts on biodiversity of using biological resources, and the reduction of adverse impacts from these and other sources through appropriate policy, law, regulation, and other means
- policies and mechanisms for governing access to and deriving equitable benefits from genetic resources
- policies and mechanisms for fostering access to and the development and adaptation of technology
- exchange of information at all levels, including among neighboring states, especially those that share common ecosystems
- technical and scientific cooperation at all levels, especially among neighboring countries
- measures to safely manage biotechnology development and the distribution of its benefits
- financial resources to cover national plans and to help other nations achieve their own goals and those of common international interest (such as the management of shared ecosystems and technological development)
- reports to the national executive, parliament, the public, and the international community.

PROVISIONS OF THE CONVENTION IN BRIEF

Box 10 provides an abbreviated list of selected Convention articles that call for action. The checklist of action points was drawn from these articles. (*For a listing of the full text of the articles, see Annex F*). The full official text of the Convention is the ultimate reference, but this brief list of substantive points should help biodiversity planners focus on the various types of action.

BOX 10

Abbreviated and Selected List of Provisions of the Biodiversity Convention to which Action Can Be Addressed

(Please refer to full text of the Convention articles in Annex F)

ARTICLE	ACTION
BASIC CONSIDERATIONS	
3	• States have the right to use their own resources pursuant to their own policies, and the responsibility to ensure that activities within their jurisdiction or control do not cause damage to the environment of other States or of areas beyond the limits of national jurisdiction
5	• cooperate with other Parties, in respect of areas beyond their own national jurisdiction, and on other matters of mutual interest
6a	• develop national strategies, plans or programmes, or adapt existing plans, to address the provisions of the Convention
6b	• integrate biodiversity work into sectoral and cross-sectoral plans, programmes and policies
IDENTIFICATION AND MONITORING	
7a-b	• inventory species, genetic materials, habitats, ecosystems and adverse impacts
7c	• manage and monitor adverse impacts on biological diversity
7d	• establish data management system for above items
***IN SITU* CONSERVATION**	
8a-b	• establish and manage a system of protected areas
8c	• manage biological resources outside protected areas
8d	• protect ecosystems and natural habitats and maintain viable populations of species in natural surroundings
8e	• manage and develop sustainably in areas adjacent to protected areas
8f	• rehabilitate and restore degraded ecosystems and promote the recovery of threatened species
8g	• control the risks associated with the use and release of living modified organisms
8h	• prevent the introduction of, control or eradicate those alien species which threaten ecosystems, habitats or species
8i	• reform policies and resource use practices to achieve compatibility among resource uses and conservation objectives
8j	• preserve and maintain knowledge, innovations and practices of indigenous and local communities and equitably share benefits
8k	• develop legislation and other regulatory provisions for the protection of threatened species and populations
8l	• regulate and manage adverse impacts
8m	• cooperate on *in situ* conservation in neighboring countries, providing financial and other support to developing countries

Continued on page 40

Continued from page 39

***EX SITU* CONSERVATION**	
9a/b	• establish and strengthen *ex situ* capabilities, especially in the country of origin of biotic components
9c	• recover and reintroduce threatened species
9d	• manage collecting procedures to avoid adverse impact on natural populations
9e	• cooperate with *ex situ* conservation facilities in developing countries
SUSTAINABLE USE OF COMPONENTS OF BIOLOGICAL DIVERSITY	
10a	• integrate biodiversity considerations into national planning and decision-making procedures
10b	• reform and manage biological resource uses to minimize adverse impacts on biodiversity
10c	• protect and encourage compatible customary uses of biological resources
10d	• support restoration of degraded areas by local residents where biodiversity has been reduced
10e	• encourage cooperation between governmental authorities and the private sector to develop sustainable use for biological resources
INCENTIVE MEASURES	
11	• adopt incentives for conservation and sustainable use
RESEARCH AND TRAINING	
12a	• establish education and training programs
12b	• encourage research on conservation and sustainable use
12c	• cooperate in the application of scientific and technological advances in conservation and use
PUBLIC AWARENESS AND EDUCATION	
13a	• promote awareness and understanding of biodiversity conservation and use
13b	• cooperate to develop educational and public awareness programmes
IMPACT ASSESSMENT AND MINIMIZING ADVERSE IMPACTS	
14a	• introduce biodiversity considerations into Environmental Impact Assessment procedures
14b	• ensure that biodiversity considerations are taken into account in policies and programmes of other sectors
14c/d	• establish institutional mechanism(s) with other countries to coordinate information and action on biodiversity-adverse activities beyond national jurisdiction
14e	• establish national and international emergency response mechanism(s) to address danger to biodiversity
ACCESS TO GENETIC RESOURCES	
15.2	• create conditions to facilitate access to genetic resources for environmentally sound uses by the Contracting Parties, and not impose restrictions that run counter to the objectives of the Convention
15.4	• establish policies as to how access may be provided based upon mutually agreed terms and subject to this Article
15.5	• establish policies as to how prior informed consent for access to genetic resources may be granted

Continued from page 40

15.6	• develop and carry out scientific research with participation of all involved parties
15.7	• establish policies, laws and administrative measures regarding the fair and equitable sharing of results and benefits gained from the research, development, and utilization of genetic resources
ACCESS TO AND TRANSFER OF TECHNOLOGY	
16.1-3	• provide and facilitate access to and transfer of technology, consistent with intellectual property rights
16.4	• establish mechanisms with the aim that the private sector facilitates access to joint development and transfer of technology
16.5	• take action to ensure that intellectual property rights are supportive of and do not run counter to the provisions of the Convention
EXCHANGE OF INFORMATION	
17.1	• facilitate exchange of publicly available information
17.2	• recognize the wide scope of information for exchange, including the results of technical, scientific, and socioeconomic research as well as information on training and surveying programs, specialized knowledge, and indigenous and traditional knowledge • facilitate the repatriation of information
TECHNICAL AND SCIENTIFIC COOPERATION	
18.1, 3-5	• facilitate and promote international technical and scientific cooperation in conservation and sustainable use • promote joint research ventures
18.2	• develop and strengthen national human and institutional capacity
HANDLING OF BIOTECHNOLOGY AND DISTRIBUTION OF ITS BENEFITS	
19.1	• take legislative, administrative, or policy measures to participate in biotechnological research
19.2	• promote access to results and benefits from biotechnologies
19.3	• prepare national perspective and participate in consideration of biosafety procedures and instruments
19.4	• provide information on genetically modified organisms
FINANCIAL RESOURCES	
20.1	• provide financial resources and incentives to support the national biodiversity programme
20.2-7	• developed country Parties: provide new and additional financial resources to other Parties to meet the objectives of the Convention, giving special consideration to the needs of the least developed countries, small island states, and vulnerable environments
FINANCIAL MECHANISM	
21.4	• strengthen existing financial institutions to provide financial resources for the conservation and sustainable use of biodiversity
REPORTS	
26	• present reports to the Conference of the Parties

5

Chapter Five • An Illustrative Biodiversity-Planning Process: The Basic Steps

In this Chapter:

- *Introduction*
- *Illustrative Biodiversity-Planning Steps in Brief*
- *Illustrative Biodiversity-Planning Steps in Detail*

An Illustrative Biodiversity-Planning Process: The Basic Steps

INTRODUCTION

A simple planning model—both preliminary and illustrative—can be drawn from the experience and plans of the cooperating countries. *(See case study profiles in Annex A and References.)* Its basic steps are consistent with general planning theory and country experience. Explicit and logically sequenced, the steps facilitate the integration of biodiversity considerations into national and sectoral planning.

ILLUSTRATIVE BIODIVERSITY-PLANNING STEPS IN BRIEF

The simplified overview shown in Box 11 contains seven major steps. The team begins its work by organizing its institutional framework and finances. Then, the team can assess the situation, consider goals and objectives, and weigh alternative actions and policies to determine which will best cover the identified gaps. The team then "operationalizes" the policies and actions to forge partnerships for implementation, set timelines, make commitments, and specify as far as possible how the plans will prompt action to further conservation, sustainability, and equity objectives. During implementation, the team assesses how well the plan's objectives are being reached, what problems have arisen, and how nature and society have evolved. This and earlier steps provide the information needed to prepare reports to government, citizens, and the international community. This information also sets the stage for revising earlier decisions in order to adapt to changing societal, institutional, and natural conditions.

ILLUSTRATIVE BIODIVERSITY-PLANNING STEPS IN DETAIL

Now, the seven steps are examined in detail and subdivided into more specific components. Again, flexibility is critical. The sub-steps should not be taken as rigid "cookbook" prescriptions, but rather as general guidelines.

STEP 1: GETTING ORGANIZED

Five sub-steps are suggested:

First, establish the institutional framework.

Who can lead? Who has the authority to prepare biodiversity plans and policies? It is essential to review the existing legislative, policy, and administrative context within the country, as well as its international obligations, to clarify the mandates and responsibilities of public agencies, private sectors, nongovernmental organizations (NGOs), and communities. Forestry laws, conservation-related treaties, environmental regulations, and trade agreements all help define for the planners the reality within which biodiversity can be analyzed. Far from arriving on an empty desk, the Convention places additional strain on people, institutions, facilities, and financial resources that are already taxed. Before launching a biodiversity plan, planners must scope out current international and national commitments, and seek orientation from various national-level plans, programs with rural and indigenous communities, and other policies that relate to biodiversity.

BOX 11

Basic Steps in an Illustrative Biodiversity Planning Process

GETTING ORGANIZED

Establish the institutional framework; designate leadership; create the participatory approach; form the interdisciplinary and inter-sectoral team; assign the budget.

ASSESSMENT

Gather and evaluate information on the status and trends of the nation's biodiversity and biological resources, laws, policies and organizations, programs, budgets, and human capacity; create a preliminary statement of goals and objectives, identify gaps and do a preliminary review of ways to close gaps; make a rough estimate of costs and benefits and unmet needs.

DEVELOPING A STRATEGY

Determine goals and operational objectives; analyze and select specific measures that will close the gaps identified in the assessment; further consult and modify until consensus is reached on acceptable targets and mechanisms; characterize stakeholders and what they can do; write a statement of the strategy, consisting of the actions and investments called for to meet the goals and objectives; at this stage consult closely with other conservation and development plans and sectors, including conducting a national dialogue with all interested stakeholders.

DEVELOPING A PLAN OF ACTION

Determine which organizations (public and private) will take charge of implementing which activities denoted in the strategy, geographically in what location or region, by what means, and with what resources (people, institutions, facilities, and funds); distinguish time phases for action.

IMPLEMENTATION

Launch activities and policies in practical terms; have partners take charge of particular elements of the plan; have biodiversity planners become "biodiversity implementors" in the key ministries, nongovernmental organizations, communities, indigenous groups, business, and industry, acting out of self, group, or business interest and commitment.

MONITORING AND EVALUATION

Establish indicators of success, determining which organization(s) will monitor which factors and the methods that will be employed; track the status and trends of biodiversity (species, genes, and habitats and landscapes); implement policies and laws; implement specific strategic action and investment; and develop capacity (people, institutions, facilities, and funding mechanisms).

REPORTING

Determine who will prepare which type of report, who will receive each, and what format, content, elements, and timetable makes sense. Types of reports include:

- annual status reports to the national chief executive, parliament, and the people
- country study
- national strategy
- action plan
- five year status report on biodiversity and biological resources
- periodic report to the Convention, the UN Sustainable Development Commission, and other international forums.

Second, determine which groups and individuals will participate in planning.

Who can be involved? When do the various interests get involved? Which organizations and institutions are expected to play key roles in implementing the Convention? How can they become involved in the strategy-planning process, lending it their experience, ensuring their agreement with the policies that will be formulated, and securing their commitment to the action plan? What mechanisms best foster their involvement? Countries have followed various approaches to participation, ranging from one that involves a full complement of government and private interests, to one in which government experts prepare draft plans that are then opened up for discussion and input. Box 12 provides guidance on nurturing political commitment for strategy planning.

According to the country cases, participation cannot be overemphasized. But which stakeholders should help prepare and implement the plan? Who will characterize the roles, perspectives, and capacities of the key interest groups in the field, the laboratory, and industry? Who will involve these groups in the steering committee, task forces, working groups, or other mechanisms of the process and clarify how agreements among these interested groups and government agencies will be reached? Rural communities, women's groups, indigenous groups, business, resource-management

BOX 12

Nurturing Political Commitment

Building support among key politicians for the biodiversity strategy and its implementation is one of the most important tasks of the biodiversity planning team. Without it, effort at all other levels can be frustrated. There are several ways to engage politicians in the process.

TARGET THE MAIN-LEVEL CONSTITUENTS

Above all, politicians are responsive to the views and demands of their local constituents:

- conduct local and district meetings to discuss the strategy
- emphasize the local development benefits of biodiversity conservation (the more specific, the better)
- initiate demonstration activities in strategic local areas early in the process

CAREFULLY PLAN A WELL-TARGETED COMMUNICATION STRATEGY

National and local politicians should be the focus of an outreach program including:

- regular briefing sessions for influential parliamentary committees and members
- high-profile workshops as an initial step to involve politicians in defining the strategy process

SEEK AND PUBLICIZE EARLY SUCCESSES

Define demonstration projects very selectively, looking for the maximum opportunity for quick and positive results. Set precise priorities and targets for action early in the process.

DEFINE PERIODIC DECISION POINTS IN THE PROCESS

Build in steps where key strategy documents require high-level consideration and approval before the process can advance.

MAKE THE BIODIVERSITY STRATEGY OFFICIAL POLICY

Decide from the start that the final strategy document will be signed as government policy. If appropriate, have the strategy considered and endorsed by the relevant bipartisan committees of parliament.

INVOLVE CABINET ON A REGULAR BASIS

Include oversight of the strategy process in the mandate of a cabinet subcommittee. Where a country has undertaken National Environmental Action Plan (NEAP), such a committee should already exist. In any case, build on existing structures.

agencies, foreign affairs personnel, researchers and universities, and NGOs that are active in development and the environment may contribute importantly to the process.

On joining the biodiversity-planning process, individual participants automatically become "biodiversity planners" and members of the team regardless of their formal or informal education, training, or background. Space at the planning and negotiating table is available to all stakeholders and interests.

Third, establish or use existing necessary institutional structures to manage biodiversity planning.

Initially, every effort should be made to take advantage of established institutional structures to avoid overlap and duplication of effort. In the cases reviewed here, most countries created one or more of the following mechanisms, generally within existing institutions:

1. ***Focal point:*** A focal point for managing the process helps provide the necessary leadership, communication, and coordination, foster broad-based participation, and establish a network of scientific and technical skills. This focal point may, variably, be a government department, a special high-level task force, or a consortium of agencies. It can be set up by the president, parliament, the planning board, a parastatal organization, an NGO, or a regional body. Many countries have set up a "national biodiversity unit" for this purpose.
2. ***Steering Committee:*** A steering committee can provide overall policy guidance for the strategy process. This group can comprise representatives of public and private groups that will play an important role in planning and implementation, including those from government agencies, academic and research institutions, nongovernmental and community-based organizations, resource-user groups, and corporate entities. Vital here is the participation of women, minorities, and indigenous peoples. Also, constituent groups without representatives in the capital city must be involved in steering committee activities.
3. ***Technical Task Forces and Working Groups:*** Special long-term or ad hoc teams with expertise in economics, biology, social science, rural community affairs, and traditional knowledge can be appointed to support the steering committee or focal-point body. Other groups can help representatives of various interests contribute to strategy-building, including the analysis of such issues as access to particular genetic resource areas, development and coordination among botanic gardens and zoos, and the determination of priorities for biodiversity inventory.

Whatever the structure, the early meetings should perhaps focus on the importance of biodiversity to the country. Why is a plan being prepared? To whom is it targeted? What is to be achieved by its implementation?

Fourth, determine the budget needed to cover the process, including personnel time, travel, equipment and supplies, reporting costs, and overhead.

Funding for planning can be provided through internal or external sources. All developed and some developing countries reported in the case studies financed their biodiversity planning through government agency budget allocations or special appropriations. Most developing countries, countries with economies in transition, and small island states, however, called on external support through the Global Environment Facility or other multilateral or bilateral sources. The catalytic role of external funding for these "getting organized" activities cannot be overstated. Through cooperative projects with outside groups, country institutions can often benefit from the methods and experience used elsewhere *if* the host country retains leadership of the project.

Fifth, if appropriate, review and study available external funding mechanisms and become familiar with any necessary funding or project-preparation guidelines.

New financial mechanisms established to implement provisions of the Convention and new guidelines developed by other funding organizations for biodiversity project support may well influence the strategy or action planning process, so countries will want to keep abreast of both.

STEP 2. ASSESSMENT

Once it is clear to all who is responsible, what their institutional mandate is, who will participate and when, and which organizational mechanisms will be used, the process can continue. Every nation possesses biotic wealth. Forest, soil, water, coastal, marine, and other environments have both wild and domesticated components. Each nation has institutions for managing biological resources, including cultural practices and norms, land-tenure systems, and public and private organizations. In addition, each nation has personnel trained in the various associated fields and familiar with both the programs already under way and the laws and policies that govern and guide conservation and use.

As noted earlier, UNEP's *Guidelines for Country Studies* (UNEP, 1993a) provides extensive orientation and guidance for preparing a systematic and comprehensive biodiversity assessment. These guidelines should be consulted in the assessment phase of the biodiversity planning process.

A country study may provide all of the necessary data and information to move forward to strategy planning. Information collected in other previous exercises may also help. It is important to assess all sources of biodiversity information to determine whether a new national biodiversity assessment is needed to minimize duplication of efforts. The planning team must determine its data and information requirements and gaps.

Three sub-steps are suggested:

First, if a country study has been completed or is in progress, analyze its findings and conclusions to determine what additional information, if any, will be needed for the strategy process.

Second, analyze all national planning efforts that address biodiversity assessment and management, such as National Environmental Action Plans, National Conservation Strategies, Tropical Forestry Action Plans, Country Environmental Studies, sector studies, etc., and integrate any relevant information with the findings of the country study for use in strategic planning.

Third, if no country study has been carried out and earlier planning efforts do not provide enough information to support the planning process, the UNEP Guidelines for Country Studies (UNEP, 1993a) should be consulted and the necessary information obtained.

ILLUSTRATIVE EXAMPLE 1

Step 1 Getting Organized: Steering Committee and Working Groups

The involvement of the broad spectrum of people and institutions with a stake in biodiversity is essential to developing an effective national strategy and action plan. The initial steering committee and working groups of the Vietnam National Biodiversity Action Plan included numerous ministries, agencies, academic institutions, and research centers:

- Ministry of Forestry
- Institute of Ecology and Biological Research
- Forest Inventory and Planning Institute
- Ministry of Agriculture
- Ministry of Science, Technology, and the Environment
- Institute of Aquatic Products
- Department of Protection of Aquatic Resources
- Mangrove Ecosystem Research Centre
- Haiphong Marine Research Centre
- Nha Trang Marine Institute
- University of Hanoi
- Centre for Resource Management and Environmental Studies
- Institute of Economic Ecology
- Council of Ministers
- State Planning Committee

Source: Case study material prepared by IUCN for this report

Developing a comprehensive database on the nation's biota, institutions, programs, and capabilities will take several years. Critical now is gathering preliminary information to support decision-makers seeking to formulate goals and operational objectives and attending to the most urgent matters of conservation and development. It may well be uneconomical and scientifically unnecessary to await the complete results of a national survey before making important strategic decisions. In many countries, the available information may justify immediate action. Longer-term inventory work will support subsequent rounds of decision-making and planning.

Several general guidelines can help planners develop an information base:

- begin the process by identifying and documenting current sources of information;
- use mutually agreed systems of classification and units of measurement for quantifying the status of and trends in basic factors; for example, use the IUCN classification of protected areas (IUCN, 1994a), taxa groups, and species status (Mace and Stuart, 1994) and use the genetic resource classifications used by the International Plant Genetic Resources Institute and the UN Food and Agriculture Organization;
- analyze each factor in terms of the stated operational objectives and in comparison with other countries in the region and the world;
- be aware that inventory and research are expensive and take time, but also that data shortages have important implications; although decisions based on ignorance are risky, waiting for extensive inventory to be created may delay critical choices and foreclose options to save important genes, species, and habitats; and
- pay special attention to institutional, personnel, facility, financial, and technological limitations.

Significantly, the assessment should go well beyond a listing of species and genetic traits and a review of government offices and programs. The assessment should also provide background information on legislative, policy, and administrative frameworks for biodiversity management in the country (including the status of local, regional, and international agreements, especially the Biodiversity Convention). It should note the mandates, instructions, and methods to be followed and the agreed-to process for preparing, approving, and implementing the strategy. It should also include background information on such issues of fundamental importance to biodiversity as access to genetic resources, ownership, property rights, relations with indigenous people, technology development, and financial arrangements.

The data and information gathered in the assessment is intended to support the biodiversity planning process as discussed in the following pages. *(See Examples 2, 3, 4, 5, 6, 7)*

STEP 3. DEVELOPING A STRATEGY

With an assessment, the national team can define goals and objectives, as well as mechanisms for covering the gaps.

Seven sub-steps are suggested:

First, articulate and debate the overall vision for conserving biodiversity and for making the use of its components more sustainable.

The purpose is to establish a perspective that reaches beyond the present and sets targets for the future, thereby calling forth expanded capacity and redoubled efforts. Initial ideas may come from national legislation and policy or from literary works, historical traditions, and current practice. The overall vision or goal will need to be discussed carefully because it can inspire and orient. Partners and stakeholders will need to "buy in" to this vision, lest they find themselves feeling excluded.

The overall aim is to launch and maintain a "biodiversity planning process."

A vision statement might read:

By 2015, the safety of the nation's biotic wealth will be ensured, its values appreciated by society at large, and the uses to which the nation puts its biological resources will be on the path to sustainable management. Human and institutional capacity and international relations will be developed so that the economic, intellectual, and cultural benefits of biodiversity can be shared equitably while international commitments are fulfilled.

Second, visions and goals for biodiversity can be broken into component parts that give pragmatic direction to action.

Team members may wish to begin with the

three objectives of the Convention shown on page 37 as points of departure for their own analysis. Additional guidance on formulating biodiversity objectives can be found in the *Global Biodiversity Strategy* (WRI/IUCN/UNEP, 1992):

- Save biodiversity by protecting genes, species, habitats, and ecosystems. This is typically accomplished by (a) maintaining key natural habitats through *in situ* measures (protected areas); (b) using *ex situ* measures (botanic and zoological gardens, gene banks, tissue culture collections, captive breeding, etc.); (c) incorporating protective measures within human-modified landscapes and waterscapes (agriculture, forestry, fisheries); and (d) restoring degraded habitats and populations of endangered species.
- Study biodiversity by fostering its understanding and appreciation. This can be achieved by (a) documenting the components, distribution, structure, and function of biodiversity through

ILLUSTRATIVE EXAMPLE 2

Step 2 Assessment: Inventory

Identifying and compiling existing data on biological resources is one of the essential tasks in the assessment phase of biodiversity planning. Data and information can be centralized from the multitude of sources both within and outside of the country. Including examples of a country's biodiversity can strengthen points and arguments. The Chinese Biodiversity Action Plan included several examples of the country's vast biodiversity.

ESTIMATED NUMBER OF ENDANGERED SPECIES IN CHINA

TAXA	NUMBER OF SPECIES	NUMBER OF ENDANGERED SPECIES	ENDANGERED SPECIES AS PERCENTAGE OF TOTAL SPECIES
Vertebrates			
Mammals	499	94	18.8
Birds	1,186	183	15.4
Reptiles	376	17	4.5
Amphibians	279	7	2.5
Fish	2,804	97	3.5
Subtotal	5,144	398	7.7
Higher Plants			
Bryophytes	2,200	28	1.3
Pteridophytes	2,600	80	3.1
Gymnosperms	200	75	37.5
Angiosperms	25,000	826	3.3
Subtotal	30,000	1,009	3.4
Total	35,144	1,431	4.1

Source: NEPA, 1994

ILLUSTRATIVE EXAMPLE 3

Step 2 Assessment:
Financial Resources and Unmet Needs for Conservation

Assessing existing financial resources for biodiversity conservation and determining the amount of unmet financial needs was the emphasis of many of the first biodiversity country studies. The Indonesian Country Study on Biodiversity determined financial resources and unmet needs for *in situ* biodiversity conservation from 1992 to 2002.

Financial resources and unmet needs for *in situ* biodiversity conservation, 1992-2002
Estimated finances per year in 1991 US$ millions

PERIOD	CURRENT FUNDS	EXPECTED FUNDS	TOTAL FUNDS	NEEDED FUNDS	UNMET FUNDS	% MET
1991-1992	12		12	190	178	6
1993-1997	12	29	41	190	147	22
1998-2002	12	58	70	190	116	37

Source: GOI, 1992

ILLUSTRATIVE EXAMPLE 4

Step 2 Assessment:
Valuation of Biodiversity

An important ingredient for the successful implementation of a strategy and action plan is the commitment of the government, business, and general public. The assessment can provide for an economic rationale of the importance of conserving biodiversity. The Norwegian Country Study on Biodiversity presented the economic value of many harvestable biological resources. (US$1=approx. 6.8NOK)

(Yield of various forms of harvesting of biological production in Norway. The yield from roe deer hunting is an estimation. The value of game meat is based on a price of 60 NOK per kg. Values in 1990-NOK.)

PRODUCTION	QUANTITY (TONNES)	VALUE (NOK)
Hunting		
Elk	4,327 tonnes	260 mill
Red deer	697 tonnes	42 mill
Wild reindeer	335 tonnes	20 mill

PRODUCTION	QUANTITY (TONNES)	VALUE (NOK)
Roe deer	25,000 individs.	21 mill
Small game	1.3 mill individs.	—
Farmed salmon (1989)	115,222 tonnes	4,344 mill
Salmon, trout and sea charr fished in the sea (1989)	496 tonnes	18 mill
in rivers (1989)	476 tonnes	18 mill
Marine fisheries (1988)	1,749,600 tonnes	5,032 mill
Forestry		
Timber felled (1988-1989)	10.7 mill m^2	3,748 mill
Agriculture		
Plant products, total value (1989)		6,081 mill
Meadow & pasture products	4,400,000 tonnes	
Cereals	1,600,000 tonnes	
Oil plants	140,000 tonnes	
Root vegetables	4,840,000 tonnes	
Other vegetables	190,000 tonnes	
Fruit	40,000 tonnes	
Domestic Animals		
Milk - cows	1 855 mill l	6,691 mill
Milk - goats	27 mill l	139 mill
Meat		
Cattle and calves	81,642 tonnes	2,825 mill
Sheep and lambs	23,693 tonnes	680 mill
Pigs	83,003 tonnes	2,200 mill
Semi-domesticated reindeer	2,622 tonnes	—
Hens, chickens, broilers	17,986 tonnes	395 mill
Other meat	3,493 tonnes	84 mill
Eggs	50,994 tonnes	674 mill
Wool	4,964 tonnes	236 mill
Honey	2,000 tonnes	63 mill
Furs	1,045,400 pelts	146 mill

Source: DNM, 1992

ILLUSTRATIVE EXAMPLE 5

Step 2 Assessment: Economic Benefits of Biodiversity

One component of a national biodiversity assessment can be the determination of the economic value of biodiversity to specific productive sectors. The Country Study of Costa Rica determines the estimated value of biodiversity both within and outside of protected areas.

	ESTIMATED ANNUAL BENEFITS DERIVED FROM BIODIVERSITY (US $)
Protected Areas	
Tourism	87,000,000
Water and Electricity	103,000,000
Biological Cataloging	No data
Unprotected Areas	
Medicine	82,000,000
Wild Plants	20,000
Hunting and Birds	144,380
Fishing	36,000,000
Forestry	97,907,258
Farming/stockbreeding	1,158,000,000

Source: MIRENEM/MNCR/INbio, 1992

ILLUSTRATIVE EXAMPLE 6

Step 2 Assessment: Costs of Biodiversity Conservation

An assessment provides the opportunity for cost-benefit analysis for the management and conservation of biodiversity. The Norwegian Country Study on Biodiversity articulated the costs to government departments for conservation of the environment and biodiversity. (US $1=approx. 6.8 NOK)

COST ENTRY	MILLIONS	1990 NOK
Ministry of the Environment, budget, total		***235***
Protecting and managing areas	50.4	
Game management	50.4	
Management of salmon and stationary freshwater fish	40.8	
Management of coastal and marine wildlife	8.4	
Interference with nature and bioengineering	16.6	
Agricultural landscape	6.3	
Environmental monitoring/Tolerance limits/ Climate monitoring	9.6	
Sum of costs particularly related to biological diversity	182.5	
Other ministries:		
Ministry of Fisheries		***165.7***
Pure environmental measures	98.7	
Measures with significant relevance for the environment	67.0	
Ministry of Agriculture (pure environmental measures)		***605.8***
Ex situ preservation	7.0	
Sum all ministries		***9,727.9***
Pure environmental measures	4,669.8	
Measures with significant relevance for the environment	5,058.1	

Source: DNM, 1992

ILLUSTRATIVE EXAMPLE 7

Step 2 Assessment: Management Costs for Parks and Protected Areas

One component of analyzing national expenditures on biodiversity could be an assessment of expenditures on protected areas. The Indonesian Biodiversity Action Plan calculates the management costs of its parks and protected areas per km^2 for a comparative analysis.

MANAGEMENT COSTS FOR PARKS AND PROTECTED AREAS (1987-1988)

PARK	AREA	ANNUAL OPERATING BUDGET (US $)	ANNUAL COSTS PER KM^2
Gunung Leuser	10,946	232,357	21
Baluran	279	187,172	671
Bali Barat	772	159,527	207
Dumoga Bone	2,780	136,874	49
Gede Pangrango	152	120,714	794
Komodo	407	111,817	275
Ujung Kulon	78	90,274	115
Kerinci Seblat	14,846	79,606	5
Tanjung Puting	3,552	68,639	19

Source: GOI, 1993a

inventory, data management, and research activities; (b) understanding the roles and functions of genes, species, and ecosystems through research and monitoring; (c) ensuring the maintenance and application of indigenous knowledge and practice; (d) understanding the complex links between modified and natural systems through research at landscape scales; and (d) building an awareness of biodiversity's values. Action can include providing opportunities for people and decision-makers to appreciate nature's variety, integrating biodiversity issues into educational curricula, and making sure that the public and decision makers have access to information on biodiversity, especially on developments that will influence it locally.

- Use biodiversity sustainably and equitably by shifting to patterns of use and consumption that don't draw down "nature's capital" and to patterns of distribution that are equitable. This means (a) nurturing biological resources so that they retain their productive capacity indefinitely, (b) making sure that biodiversity is used to improve the human condition, and (c) seeing that the economic, cultural, and other (monetary and nonmonetary) costs and benefits from these resources are shared equitably.

Objectives will subdivide the goal into clusters of specific operational-level action points. Some guidelines for drafting objectives suggest that they should:

- be comprehensive and reflect various national

aspirations for conservation and the sustainable use of biodiversity;

- be defined early in the planning process by empowered representatives of the broad range of stakeholders; naturally, the objective statements, along with other elements of the strategy, will be subjected to review and revision following subsequent consultations with stakeholders;
- build on existing work achievements and ongoing programs and agreements;
- be narrow and focus national efforts on explicit targets; and
- be consistent with the goals and objectives stated in national laws and policies and with international agreements.

Third, analyze the status and trends from the assessment (step 2) and compare the actual situation with the related target or objective.

What are the unmet needs? What opportunities can be seized and what instruments can be used to close these gaps? What obstacles lie in the path? What basic issues need to be addressed?

The inventory of status and trends should help determine the types of challenges, constraints, limitations, and obstacles that should be resolved. Additionally, team members will identify opportunities for progress that have not been fully appreciated and can be implemented either easily or with minimal action.

Fourth, formulate options for action that cover the most important gaps or that address pressing problems, issues, and undeveloped opportunities.

Drawing distinctions between realities and objectives is the heart of strategic planning. Various options can be considered to help close each gap. These include training and other means of building human capacity, constructing facilities, establishing national environment funds, shifting taxation policies, and creating incentive programs to foster private investment in new technologies and industrial processes.

Each option should be measured against social, economic, and ecological values. Methods for quantitatively analyzing the merits of options based on these values are currently underdeveloped, however, simply examining the positive and negative features of each option can provide decision-makers with a reasonable basis for decision. Particular attention should be given to ethical and cultural values and to the economic costs and benefits involved. Open debate with stakeholders to gain from their insights and hear their perceptions can help. Such dialogue also ensures that the measures designed and selected are realistic and can, in fact, be implemented by one or more groups.

Fifth, assess and develop mechanisms for technology cooperation and financial assistance.

- Assess and establish appropriate measures for facilitating technology cooperation and information sharing, as required in Article 16. Take into consideration developing countries' needs for particular types of technology and information, including those related to planning, negotiation, data and information management, captive breeding, plant and animal reintroduction, inventory, biotechnology, etc.
- Developed countries will assess and establish appropriate means for the provision of additional financial resources to enable developing countries to meet the additional costs of implementing the Convention, as required in Article 20.

Sixth, establish criteria and priorities to help choose from among options. Guidelines include:

- identify the options most likely to succeed;
- identify the options that can be implemented given the available personnel, institutions, facilities, funding, and time frame—or that could expand these means;
- identify the options that promise the sharing of costs and benefits of biodiversity and biological resource use will be equitable;
- identify the options that promise to provide the most or best outcome for a given amount of funding; alternatively, note the options that provide the same level of outcome for the least cost;
- identify those options that would provide greater positive impact on achieving sustainable livelihoods and capacity building, including technology sharing; and,
- identify options with outcomes that promise to meet the goals and objectives of the Convention,

Agenda 21, other international agreements, and national laws and policies; meet social, economic, and ecological standards in the interests of local communities, including equitable sharing of benefits; and can be adequately integrated into the policies and practices of other sectors.

After this first evaluation, those options that meet criteria similar to those discussed above can then be ranked into further categories, for example:

- highest priority: actions that address urgent matters, inequities to constituencies, and policy inconsistencies;
- second priority: actions that reform resource use, institutions, and legislation; and
- third priority: actions that develop and expand national capacities and that reform education programs.

Seventh, match actions and objectives.

With these decisions made, the program may choose to publish a national strategy that focuses on the goals and objectives and the selected policies and actions that will cover the identified gaps. A national strategy responds to the question: Of all that could be done in biodiversity conservation and development to help the nation meet its vision and goals, which tasks are strategically most important and feasible? *(See Examples 8-19.)*

STEP 4. DEVELOPING A PLAN OF ACTION

Six sub-steps are suggested:

First, the actions identified in Step 3 can be broken down into discrete types of tasks that make it easier to link related sectors (forestry, agriculture, fishing, education, tourism, etc.) and interest groups (rural, community, indigenous, development and environment NGO, biotechnology industry, women's, etc.) to the actions to be taken.

For example, tasks related to the promotion of national biotechnology capacity will, in some countries, perhaps be taken up by the national science and technology council, the ministry of commerce, the national university departments of phytochemistry and engineering, private pharmaceutical producers and agricultural research groups, and consortia preparing draft policies and legislation on biosafety. Tasks related to the sustainable use of forest resources will perhaps fall to the university, the private sector, and the forest service, which—with local communities, rural NGOs, and private marketing and distribution firms—can conduct and promote sustainability research, new technology development for forest management and timber harvesting, local community-based forest-management schemes, and exploration of new nontimber forest products.

Second, the roles and responsibilities of each cooperating (public and private) institution can be described by asking:

- Who will implement each of the proposed tasks? The question of participation in biodiversity planning will already have been discussed by the team before these planning steps are initiated. It will have determined who will be involved, at what stage, and through what mechanisms. Here the question becomes: Which institutions and organizations (public agencies of central and local governments, local communities, indigenous groups, business and industry, university, women's groups, and NGOs) will accept the responsibility for implementing each task? And how will these arrangements be made?

Review of the cases available to date suggest that where partners are fully involved from the beginning of the planning process, the capabilities and interests of each partner will already have been assessed, making it easier to link partners and area of interests and responsibilities. Presumably, participation reduces the need to impose roles and responsibilities through political pressure or regulatory mechanisms and minimizes administrative costs. Self-interested motivation is also maximized.

Third, specify with partners:

- the geographic area where their contribution will be made (in their community, bureaucratic region, field office, state, or county);
- the tools and approaches to be used, including capable people, institutional cooperation, facilities, and funds.

Detailing these agreements may entail a complex series of legislative initiatives, inter-institutional meetings, multi-donor meetings, and roundtable discussions.

ILLUSTRATIVE EXAMPLE 8

Step 3 Developing a Strategy: National Vision, Guiding Principles, and Goals

An effective strategy begins with a clear vision statement and a set of guiding principles. These components can be used to identify goals and subsequent actions which are focused and consistent. The draft Canadian Biodiversity Strategy clearly articulates its vision, principles, and overall goals. Included below are the objectives for the strategy's goal of conserving and sustainably using biological resources.

A VISION FOR CANADA

A society which values all life, which takes no more from nature than nature can replenish and which leaves a world rich in biodiversity for future generations.

GUIDING PRINCIPLES

- All life forms have intrinsic value.
- All Canadians depend on biodiversity and have a responsibility to contribute to biodiversity conservation and to use biological resources sustainably.
- All Canadians should be provided with opportunities to understand and appreciate biodiversity and participate in resource and air and water and land-use decisions that affect biodiversity.
- An ecological approach to resource management is central to achieving biodiversity conservation and the sustainable use of biological resources.
- Development must be ecologically and economically sustainable.
- Biodiversity is best conserved in the wild *(in situ)*.
- The knowledge, innovations and practices of indigenous and local communities should be respected, preserved, maintained, and used with the approval and involvement of those who possess this knowledge.
- Conservation of biodiversity should proceed on the basis of the best knowledge available, using approaches that can be refined as new knowledge is gained.
- Biodiversity conservation requires global cooperative action and a sharing of knowledge, costs, and benefits.

OVERALL GOALS

- enhance efforts to conserve biodiversity where it exists naturally and to sustainably use our biological resources, particularly in resource-based economic sectors such as agriculture, forestry and fisheries;
- make the transition to an ecological approach to management that is built on a better understanding of how ecosystems function, comprehensive and reliable biological inventories, information sharing, greater recognition and use of traditional knowledge and practices, and integrated planning and monitoring systems;
- improve Canadians' understanding of the value of biodiversity and provide citizens with opportunities to contribute to the development and implementation of land and resource use policies, plans and programs;
- ensure an appropriate mix of legislation and incentives in order to encourage biodiversity-sensitive behaviour; and
- contribute to biodiversity conservation and sustainable use efforts worldwide.

GOAL 1

Conserve biodiversity and sustainably use biological resources.

OBJECTIVES

Canada's approach to conserving biodiversity and sustainably using biological resources has six key elements.

- maintain wild flora and fauna populations across Canada's diverse ecosystems, landscapes and waterscapes;
- establish networks of protected areas to conserve or sustainably use ecosystems, species, and genetic diversity;
- restore degraded ecosystems where practical and where restoration will make a significant contribution to the conservation and sustainable use of biodiversity;
- develop and implement biodiversity conservation and sustainable use policies, plans, and programs within specific sectors that use biological resources;
- develop and implement policies and programs that are aimed at preventing or reducing human-caused atmospheric changes that threaten biodiversity; and
- ensure that development and use of non-renewable resources does not result in the decline of biodiversity.

Source: FTP Biodiversity Working Group, 1994

ILLUSTRATIVE EXAMPLE 9

Step 3 Developing a Strategy: Goals

The draft Philippine National Biodiversity Strategy formulated the following goals.

GOALS

- To develop and implement a holistic and comprehensive national program for the conservation of biological diversity and the sustainable use of its components.
- To coordinate the planning and implementation of a biodiversity conservation program by ensuring that relevant activities harmonize with those of other government and non-government organizations, private sector, religious groups, communities and other organizations.
- To institutionalize the practice of biodiversity conservation and the sustainable use of resources through legislative, administrative, fiscal and other regulatory measures.
- To promote public education and understanding of the values and benefits of biodiversity conservation and of the merits of sustainable development.
- To enhance capacity building through formal and non-formal education, training, research, and institutional strengthening.

Source: PAWB/DENR, undated

ILLUSTRATIVE EXAMPLE 10

Step 3 Developing a Strategy: Objectives

The National Biodiversity Action Plan of China developed and articulated the following objectives.

OBJECTIVES

The overall objective of biodiversity conservation in China is to set in place, as soon as possible, measures for avoiding further damage, and, over the long term, for mitigating or reversing the damage already done.

Effective biodiversity conservation can achieve this objective in two ways: first, by the total protection of rare and endangered species and ecosystems facing extinction (for example, by banning their use for a period of time); and second, by the rational and sustainable use of those biodiversity resources that exist in sufficient quantities to allow for such exploitation. Because of the urgency of the threat to China's natural resources, the Biodiversity Conservation Action Plan concentrates most of its efforts on the first of these objectives through the following means:

- *in-situ* conservation in nature reserves, parks, and other protected areas;
- *in-situ* conservation outside of nature reserves, parks, and other protected areas;
- development of priorities among species for direct protection; and, in combination with analyses of the above proposals, development of options for *ex-situ* conservation in zoos, botanical gardens, aquaria, gene banks and breeding centers;
- establishment of a nationwide information and monitoring network to track the status and trends of China's biodiversity; and
- integration of conservation issues into the central economic planning of the country as a whole.

In-situ conservation is the core of the program, and is complemented by *ex-situ* means as indicated above.

Source: NEPA, 1994

ILLUSTRATIVE EXAMPLE 11

Step 3 Developing a Strategy: Principles

Principles are an important tool which can be used to guide the selection of options and actions. Australia developed the following principles to guide their biodiversity strategy process.

PRINCIPLES

The following principles have been adopted as a basis for the Strategy's objectives and actions and should be used as a guide for implementation.

- Biological diversity is best conserved *in-situ.*
- Although all levels of government have clear responsibility, the cooperation of conservation groups, resource users, indigenous peoples, and the community in general is critical to the conservation of biological diversity.
- It is vital to anticipate, prevent and attack at source the causes of significant reduction or loss of biological diversity.
- Processes for and decisions about the allocation and use of Australia's resources should be efficient, equitable and transparent.
- Lack of full knowledge should not be an excuse for postponing action to conserve biological diversity.
- The conservation of Australia's biological diversity is affected by international activities and requires actions extending beyond Australia's national jurisdiction.
- Australians operating beyond our national jurisdiction should respect the principles of conservation and ecologically sustainable use of biological diversity and act in accordance with any relevant national or international laws.
- Central to the conservation of Australia's biological diversity is the establishment of a comprehensive, representative and adequate system of ecologically viable protected areas integrated with the sympathetic management of all other areas, including agricultural and other resource production systems.
- The close, traditional association of Australia's indigenous people with components of biological diversity should be recognized, as should the desirability of equitably sharing benefits that arise from the innovative use of traditional knowledge of biological diversity.

Source: ANZECC, undated

ILLUSTRATIVE EXAMPLE 12

Step 3 Developing a Strategy: Operational Objectives and Targets for Key Habitats and Species

Setting appropriate and attainable targets in an action plan can contribute to focused activities which can be monitored for effectiveness. The Biodiversity Action Plan of the United Kingdom illustrated several examples of specific targets for key habitats and species.

EXAMPLES OF SPECIFIC TARGETS FOR KEY HABITATS AND SPECIES

HABITATS

Caledonian Pinewoods

Caledonian pinewoods consist of a mixture of mature Scots pine and birch woodland with a rich understorey of shrubs, the presence of dead and rotting wood and some open areas. The level of natural regeneration of canopy and shrub species must be sufficient to ensure the maintenance of the habitat. The presence and numbers of various birds, insects and flowering plants, ferns, bryophytes and lichens are important indicators of habitat quality.

Performance Indicator

Maintain and manage, where necessary, all existing Caledonian pinewoods (12,500 ha) and produce the correct conditions during the next 4 years to begin the process of regeneration of a further 5,000 ha.

Lowland Heathland

Lowland heathland is a range of habitats characterized by plants such as heather and cross-leaved heath found below about 250m. It supports a number of rare plant species such as the marsh gentian, Dartford warbler and sand lizard. Only one sixth of the area of the lowland heath present in 1800 now remains, representing less than 0.3 percent of England's land surface. This nevertheless represents an important proportion of the international total. It is now highly fragmented and often threatened by a lack of management and development.

Performance Indicator

Maintain, and improve by management, all existing lowland heathland (58,000 ha) and produce conditions during the next ten years to begin the process of heathland re-establishment of a further 6,000 ha in Dorset, Hampshire, Surrey, Devon, Suffolk and Norfolk.

The aims of re-establishment should be: to increase the total heathland area; to increase the heathland patch size; to link heathland patches.

SPECIES GLOBALLY THREATENED

Kite

A globally threatened species found in parts of Wales and recently successfully re-introduced into England and Scotland.

Performance Indicator

To maintain the annual red kite population increase in Wales at more than 5 percent per annum, which should result in 120 breeding pairs by 1997.

In the longer term see the species re-established throughout its former range.

ENDEMIC

Scots Primrose

A scarce endemic confined to parts of northern Scotland.

Performance Indicator

Every effort should be made to maintain the present distribution of the species at the 15 sites in Orkney and 26 in Caithness and Sutherland.

INTERNATIONALLY IMPORTANT

Gannet

A widespread but localised breeding bird found in internationally important numbers in Britain.

Performance Indicator

Maintain UK population at 160,000 pairs. Maintain population at current levels in the 14 well established colonies in Britain.

RED DATA BOOK SPECIES (DECLINING)

Bittern

A rare and declining resident, confined almost entirely to lowland marshes dominated by Phragmites needs. In winter the population is supplemented by birds from elsewhere in Europe.

Performance Indicator

Arrest the decline and maintain at least 20 booming males within the present area of distribution. Seek to increase numbers by creating suitable large reed beds in England and aim for a population of 50 pairs by 2005 and 100 by 2020.

Stone Curlew

A rare breeding summer visitor to some natural grasslands in southeast England with a declining population of less than 160 pairs. Its survival depends on the correct management of grazing in its breeding grounds and the protection of nests from farming operations.

Source: Department of Environment (UK), 1994a

ILLUSTRATIVE EXAMPLE 13

Step 3 Developing a Strategy: Components of a Strategy or Action Plan

The National Biodiversity Action Plan of Chile is divided into eight action areas.

The National Biodiversity Action Plan is composed of eight programs or sub-components, grouped in three areas of action for biological diversity:

ACTIONS FOR THE MANAGEMENT OF BIODIVERSITY

- A national program oriented toward identifying administrative divisions, clarifying capacities, and strengthening public institutions.
- A national program to develop legislation, ideally through a specific Biodiveristy Law, and regulations for natural resource protection, introduction of exotic species and marketing of native species.
- A national program proposing the incorporation of the concept of biodiversity into formal and informal education, which would strengthen the didactic material on the subject.

TECHNICAL ACTIONS IN BIODIVERSITY

- A national program to assist initiatives for identifying ecosystems which are not represented in the National System of State Protected Areas (SNASPE), to strengthen the creation of marine reserves, and to carry out a study of the populations present in the current areas of the SNASPE.
- A national program to make recommendations on the use and planning for biotic resources present in marine, lacustrine, forest, and agricultural systems.
- A national program to carry out a study on the populations present in insular ecosystems, design management plans for endemic species populations, and formulate a proposal to protect these populations from contamination.

RESEARCH ACTIONS IN BIODIVERSITY

- A national program to study conditions and strengthen taxonomic and systemic research, form a monitoring center, and stimulate scientific investigation of biodiversity and its relation to global warming and the shrinking ozone.
- A national program in biotechnology, that evaluates the economic value of native genetic resources, coordinates research in genetic engineering related to native species, and revives native folk knowledge of the uses, such as in medicines, of these species.

Source: CONAMA, 1993

ILLUSTRATIVE EXAMPLE 14

Step 3 Developing a Strategy: Components of a Biodiversity Program

A strategy and action plan can identify a series of core components to be implemented under a nation-wide biodiversity program. The draft Biodiversity Action Plan of Vietnam identified several priority action areas for a biodiversity program with specific actions to be implemented within various sector programs.

MAIN COMPONENTS OF A NATIONAL BIODIVERSITY PROGRAM IN VIETNAM

The Biodiversity Action Plan organizes various actions under appropriate sectoral programs and elaborates on the specific responsibilities, costs and timings for such actions. The major components of the program to conserve biological diversity in Vietnam are:

- To set aside a total of at least 6 percent of the land area as natural reserves representing viable examples of all major ecosystems in the country, including riverine, wetland, and marine systems.
- Maintain a forest cover of at least 40 percent, at least half of which should be natural forest, which will protect the hydrology of the country. Success will depend on the progress of the current program to sedentarize shifting agriculturalists and also to increase forest cover through natural regeneration and reforestation. It will also be important to improve the control of forest fires.
- Establish controls to ensure that biological resources are utilized sustainably and follow the principles of maximum sustainable yield.
- Ensure that illegal harvesting and trade of wildlife is brought under control, for example, Vietnam should join CITES.
- Control the emission of harmful pollution into both natural and human ecosystems through firm regulation and application of EIA (environmental impact assessment).
- Actively ensure the preservation of the full range of domesticated varieties of plants and animals in Vietnam.
- Improve capacity for management of wild and captive populations of plants and animals based on scientific research.
- Create an effective monitoring and data management system that will evaluate the status of species and habitats and identify areas of priority for protection or improved management.
- Establish a greater account of natural resource values and ecological functions in the cost of development programs and projects.
- Increase levels of international cooperation and support.

A total of 52 project concepts have been developed to tackle these actions. The total cost of these projects is $876 million.

Source: BAPPT, 1993

ILLUSTRATIVE EXAMPLE 15

Step 3 Developing a Strategy: Key Elements for Initial Action

Initiating action quickly once political commitment is secured is important in order to maintain momentum for the overall strategy process. Identifying the key first steps needed to launch a biodiversity program will contribute to expediency. The draft Biodiversity Action Plan of Vietnam proposes five key recommendations to give an immediate start to a National Biodiversity Program:

PRINCIPAL RECOMMENDATIONS

- Establish a national biodiversity authority.
- Launch a major national public awareness campaign.
- Establish a national biodiversity information system.
- Increase resources devoted to biodiversity conservation.
- Greatly expand and strengthen the protected areas system of Vietnam.

Source: BAPPT, 1993

ILLUSTRATIVE EXAMPLE 16

Step 3 Developing a Strategy: Criteria for Determining Biodiversity Significance

The National Biodiversity Action Plan of China developed the following criteria for determining biodiversity significance and conservation priorities.

CRITERIA FOR DETERMINING BIODIVERSITY SIGNIFICANCE AND CONSERVATION PRIORITY OF AREAS

Until the review called for under Objective 1 is completed, priority areas for conservation can be identified using the criteria described below:

- Areas rich in biological diversity (i.e., with high numbers of species and ecosystems);
- Areas that have a high level of endemism (i.e., high numbers of endemic species);
- Areas that are outstandingly representative of the same type elsewhere in the country, or areas that are the best representative of that type;
- Areas that are intact (i.e., relatively unmodified by human actions), and can be used as a baseline for monitoring environmental changes;
- Areas that include particularly valuable or important species; and
- Areas that are of critical ecological value, for example, as an essential migratory route, an essential breeding area, an essential feeding area, or areas essential to other important ecosystems (e.g., a water catchment for a significant body of water or wetland). These are areas whose loss will have a particularly important impact on surrounding areas or on national biodiversity conservation.

Source: NEPA, 1994

ILLUSTRATIVE EXAMPLE 17

Step 3 Developing a Strategy: Criteria for Determining Priorities for Actions

The National Biodiversity Action Plan of China developed the following criteria for determining priorities for action.

CRITERIA FOR DETERMINING GLOBAL BIODIVERSITY SIGNIFICANCE AND CONSERVATION PRIORITY OF SPECIES

GENETIC SIGNIFICANCE	Is the species unique? Is it rare or endemic? Does it have special genetic or scientific importance (e.g., the only species in a family or genus, a relic species or genus)? Is it a typical representative species or a sample of a particular species (individual)? Is it very rare or under severe threat?
ECOLOGICAL SIGNIFICANCE	Is the species a "keystone" species (i.e., one on which the health or survival of the ecosystem depends)? Does it have an especially significant impact on the ecosystem in other ways (e.g., as would a top carnivore or elephant)? Is it an especially significant ecological phenomenon, or a component of one (e.g., a complex ecological system or a complex migratory community)? Is it an important indicator species?
SOCIAL AND ECONOMIC SIGNIFICANCE	Is it a source of medicine? Does it have other economic value? Does it have outstanding cultural or historical significance? Does it have important scientific research value? Is it of significance for maintaining conditions necessary for human welfare (e.g., in pollination or pest control)?
DEGREE OF THREAT AND LEVEL OF VULNERABILITY	These two criteria must be considered together, because a species can be vulnerable but not under immediate threat, while another could be both vulnerable and in imminent danger of extinction. The priority for conservation should be given to the species that are significant (according to the criteria above) and also vulnerable and under threat. In order to determine the degree of threat and vulnerability, the following questions should be asked: How immediate, and therefore urgent, is the threat? Is the population very small? Are its numbers being rapidly depleted? Is the species or population at or close to the minimum size for survival? Is there immediate or imminent demand for it or for its habitat (e.g., because of hunting pressure or demand for land)? Is its continued viability affected by changes in human population, diversion of needed water, loss of source of food, desertification, or other major changes in conditions that affect the viability of the species? Is it readily accessible (e.g., to hunting or to other human activities that would threaten it)? Is the protection for it weak or non-existent? Will the threatened loss be irreversible?

Source: NEPA, 1994

ILLUSTRATIVE EXAMPLE 18

Step 3 Developing a Strategy: Traditional Knowledge

Strategies and action plans can benefit from structuring action around specific articles of the Convention. The draft Canadian Biodiversity Strategy utilizes this approach throughout, and includes Article 10.c, which concerns traditional knowledge. The draft Strategy also includes examples of existing efforts for illustration.

PROTECTING TRADITIONAL PRACTICES ARTICLE 10.C

protect and encourage customary use of biological resources in accordance with traditional cultural practices that are compatible with conservation and sustainable use requirements

To respond to the Convention, it is proposed that the federal, provincial and territorial governments, in collaboration with affected and interested groups and individuals:

Identify mechanisms to use traditional knowledge, innovations, and practices with the approval and involvement of the holders of such knowledge, innovations and practices, and encourage the equitable sharing of benefits arising from the utilization of such knowledge, innovations and practices.

THE BEVERLY-QAMANIRJUAQ CARIBOU MANAGEMENT BOARD

The Beverly Qamanirjuaq Caribou Management Board was set up in the Northwest Territories in 1982 to provide a mechanism to incorporate the knowledge and values held by traditional caribou users into decision-making. The Board consists of representatives from eight user communities and five government jurisdictions. The success of the Board represents a practical example of the capability of co-management structures to integrate specialized knowledge held by traditional resource users with the scientific knowledge of biologists, leading to effective, efficient, and fair management of the resource.

Source: FPT Biodiversity Working Group, 1994

ILLUSTRATIVE EXAMPLE 19

Step 3 Developing a Strategy: Indigenous Peoples' Participation

Indigenous perspectives and input can be ensured in the development and implementation of the strategy by taking specific steps and establishing adequate mechanisms. Canada's draft Biodiversity Strategy proposed the following actions for indigenous participation.

Federal, provincial, and territorial governments and indigenous communities will work together to develop an interpretation of the Convention and of traditional knowledge which connects to indigenous people and has significance to them. It is proposed that:

Indigenous communities will develop an approach to implementing the Convention with a view to reflecting distinct indigenous values, social networks, traditional economies and cultures of First Nations. This approach should:

- build upon the current networking process of meetings, workshops, and other consultations which enable indigenous communities and indigenous resource management groups to determine how they will contribute to the implementation of the Biodiversity Convention;
- demonstrate the role of indigenous knowledge and management in the conservation and sustainable use of biodiversity;
- lead to the development of community-based regimes, designed to preserve indigenous knowledge, innovations and practices at source, recognizing their potential economic and scientific value;
- establish linkages with federal, provincial, and territorial agencies that are responsible for implementing the Convention; and
- facilitate maintenance of indigenous cultural traditions which will support the communication of traditional knowledge and use of biological resources between generations and communities.

Development of an indigenous community analysis of the Convention reference to "knowledge, innovations, and practices of indigenous communities" will be encouraged, taking into account issues of intellectual property rights and the use of genetic resources. This analysis should emphasize ways in which knowledge and practices are applied in biodiversity conservation and examine how innovations can be protected by intellectual property rights.

Ways in which indigenous groups can share their knowledge and experience and develop joint programs with indigenous groups outside Canada will be examined.

Federal, provincial, and territorial governments will continue to collaborate with indigenous communities through a variety of mechanisms, such as land claims agreements, management boards, model forest programs, and other means, to conserve and sustainably use biodiversity.

Source: FPT Biodiversity Working Group, 1994

Fourth, these decisions can be recorded on a table similar to that suggested in the sample work sheet in Box 13.

Fifth, set a schedule for completing each task.

The tasks can be implemented sequentially to reflect the priorities noted in Step 3. Various factors will affect the decision about when to implement each particular task. In actual practice, of course, some flexibility is required.

These factors include:

1. ***Timing.*** Some tasks may warrant immediate intervention. For example, a site of high biodiversity may be in line for early conversion from closed forest to agriculture. Quick action may enable the protected areas agency to discuss this case with other agencies and interests, alter the plan, and protect that site before biodiversity

BOX 13

Sample Work Sheet. Record of Responsibilities for Implementing the Biodiversity Plan

(Adapted from the draft Canadian Biodiversity Strategy)

RESPONSIBILITIES FOR IMPLEMENTATION OF THE NATIONAL BIODIVERSITY STRATEGY AND BIODIVERSITY ACTION PLANS

	ORGANIZATIONS				
ACTIONS	**Government Agencies Organizations**	**Provincial Agencies**	**Regional Management**	**Farmers Association**	**Etc**
Article 3: Principle (Stockholm Principle 21)					
Ensure Canadian activities do not harm areas beyond national jurisdiction					
Article 5: Cooperation					
Cooperate with other countries for the conservation of biodiversity beyond national jurisdiction					
Article 6: General Measures					
(a) Develop a Canadian biodiversity strategy					
Agenda 21: Carry out country studies on biodiversity conservation: ascribe values to biological resources					
(b) Integrate biodiversity conservation into sectoral and cross-sectoral activities					
Article 7: Identification and Monitoring					
(a) Identify ecosystems, species and genomes important for conservation and sustainable use					
(b) Monitor components identified in (a) to determine priorities					
(c) Identify and monitor activities that may be harmful to biodiversity					
(d) Maintain and organize data from (a), (b), (c)					
Agenda 21: Collect, assess and make available relevant information in a forum suitable for decision making at all levels					
Article 8: In-situ Conservation					
Establish a system of protected areas					

assets are lost. Similarly, a place highly valued by a rural community for its medicinal plants may be slated for timber harvesting by the competent public agency. Through quick analysis and negotiations, the appropriate offices and representatives may be able to shift logging plans and have that site reassigned for community co-management.

2. ***Technical.*** Some activities may logically follow others. Assessment comes before consideration of gaps and options for action. A lack in formation may delay decision to a later stage.
3. ***Social.*** Pressing social issues, the livelihoods of rural people dependent on biological resources, confrontations between biodiversity imperatives and the desires or claims of rural residents in the area may all warrant early attention, superseding other factors that can wait several years.
4. ***Economic.*** Generating new income for local people, business, and the public treasury will always be considered vital, but some activities may well breed further impoverishment (the "decapitalization" of biological resources and biodiversity resulting in a loss of options) while others enhance biological wealth. Some investments will yield early monetary returns; others may lead to non-monetary environmental services. Which comes first depends on how much importance is assigned to them in the overall flow of activities and benefits.
5. ***Personnel, institutions, facilities and finance limitation.*** Ultimately, action will need to be scheduled carefully to make the best use of capable people, institutions, and facilities. When will the individuals with the needed skills be available and for how long? Be it a phytochemical engineer to address chemical prospecting tasks or a shaman to guide field inventory in a remote area, these scarce resources have to be carefully conserved. The support of the forest service, the university, the data center, and other institutions, and their facilities (laboratory, inventory, data-management, taxonomic, and analytical), will determine the scheduling of much of the work. And, finally, funding to enable each group to act is critical. National appropriations, special project funds, internal community or corporate sources, or international assistance all have to be obtained and programmed over time to coincide with work flow.

Sixth, plan, obtain, and distribute the budget.

Action to implement the strategy will require personnel, facility, institutional, and financial

BOX 14

Some Potential Sources of International Support for Biodiversity Strategies and Action Plans

- Biodiversity Convention
- Global Environment Facility (UNDP/UNEP/IBRD)
- World Bank
- United Nations Development Programme
- United Nations Environment Programme
- Food and Agriculture Organization of the United Nations
- United Nations Educational, Scientific and Cultural Organization
- Inter American Development Bank
- Asian Development Bank
- African Development Bank
- Bilateral aid agencies
- Private foundations
- Nongovernmental organizations, including particularly the World Wide Fund for Nature and its global network of national organizations, and the World Conservation Union (IUCN).

inputs. Each of the four categories of input can be considered separately. Thus, the team will relate the backgrounds and skills of contributors to the tasks at hand. A variety of institutional capabilities will need to be fully involved: land-tenure studies, mapping and geographic information systems (GIS), social analysis, data management, etc. The team will analyze the physical facilities needed in terms of transport, laboratories, buildings, field equipment, etc.—some of which can be met by "in-kind" contributions (without direct financial cost to the plan), and some of which will require direct financial obligation. What actual monetary commitments will be needed? The team must decide.

One good way to foster implementation and ensure the long-term viability of the process is to work out co-management and co-financing arrangements among governmental agencies and other local, national, and international interested parties. An enormous variety of possibilities can be explored for possible application, as listed in Box 14. The advent of "national conservation funds" is particularly encouraging. (IUCN, 1994b; IUCN, 1994c.) *(See Examples 20-24.)*

STEP 5. IMPLEMENTATION

At this point in biodiversity planning, the plans are translated into action. Although an enormous amount of work has been implemented on the components of biodiversity management in virtually all countries of the world, the implementation of actual biodiversity plans as comprehensive programs is only now beginning. As a result, no detailed guidelines are offered here; however, several key points emerge from this study.

First, the step of implementation is inseparable from the other planning steps that precede and follow it.

What was decided is now acted upon. What has been done is evaluated so that it can experience and guide future choices.

Second, planners (as defined in this study) become implementors.

Those individuals have self, business, community, and agency interests in not only making choices, but also in seeing that these decisions are acted upon and that results are realized and measured. It is fundamental that planners are not a class apart from those who take action and have to live with the results.

Third, the planner-cum-implementor makes the "biodiversity partnership" come alive and take on meaning.

Well beyond theory and abstraction, biodiversity management affects people, their livelihoods, their share of nature's assets, and the long-term productivity of the soils, waters, land, and seascapes.

STEP 6. MONITORING AND EVALUATION

The biodiversity-planning process features monitoring and evaluation for two reasons. First, change is inevitable and eternal—a plan in one period may be inadequate basis for a decision in another. The status of and trends in species populations, recovery of wild communities, viability of *ex situ* collections, livelihoods of rural peoples, etc., change continuously. Second, in implementing the plan, it is important to learn from each activity, including errors and successes.

To be able to monitor or track properly:

First, establish indicators that can be measured to show change in environmental factors, shifts in capacity (people, institutions, facilities, and funding), and success or failure of action.

Initial work on indicators demonstrates the feasibility of formulating practical ways to follow and qualify these phenomena (Reid et al., 1993a). *(See Box 15). See* Guidelines for Country Studies, Technical Annex 7 (UNEP, 1993a) for a full accounting of guidelines for monitoring.

Second, all relevant parties must agree on who will monitor and evaluate, and how.

Monitoring and evaluation is a sensitive topic. The choice of organization and methodology must be made openly and cooperatively. The full array of stakeholders will need to be involved in this debate, particularly because monitoring and evaluating the implementation of the various tasks maintains interest in and commitment to the plan.

Evaluation of the projects takes place at this point and throughout the planning process. How close have we come to closing the gaps? Is bio-

ILLUSTRATIVE EXAMPLE 20

Step 4 Developing an Action Plan: Priority Actions and Time Frames

When developing actions, it is important to establish implementation targets. The draft Australian Biodiversity Strategy identified priority actions and set target dates for their implementation.

OBJECTIVE

Implement the Strategy through priority actions within established time frames

PRIORITY ACTIONS

A broad range of human endeavors and natural phenomena affect the future of Australia's biological diversity and the maintenance of essential ecological processes and systems. This reality is reflected by the large number of objectives and actions in this Strategy. The objectives and their actions do not contribute equally to ensuring protection of biological diversity, nor are they equally urgent. Many of the objectives, such as those associated with ecologically sustainable development, are being pursued as part of other national strategies or initiatives. Many of the actions are being pursued and will continue to be undertaken without an urgent need for enhanced resourcing from governments. These objectives and actions will provide a guide for determining priorities for expenditure from research funds and private sources, and for community action. Those additional actions deemed to be urgent and having the capacity to make major contributions to the protection and ecologically sustainable use of Australia's biological diversity will be implemented as quickly as possible.

The priority areas for action, as depicted by their specific outcomes, are listed under Action 7.1.1 along with the time frames during which substantive results are to be achieved. These results are broadly defined and may encompass more than one of the Strategy's actions. The Strategy will be reviewed at five-year intervals to allow for assessment of progress, evaluation of priorities and, where necessary, adjustment.

ACTIONS

Priorities and Time Frames

By the year 2000 Australia will have:

- completed the identification of its biogeographical regions;
- implemented cooperative ethnobiological programs where Aboriginal and Torres Strait Islander peoples find them appropriate, to record and ensure the continuity of ethnobiological knowledge within Australia's jurisdiction, resulting in social and economic benefits to Aboriginal and Torres Strait Islander peoples;
- completed the identification and description of major ecosystems in each biogeographic region and developed specific priorities for conservation;
- established mechanisms for resourcing the development and implementation of programs and plans for the continuing management of Australia's biological diversity on public and private lands, including lands managed by Aboriginal and Torres Strait Islander peoples;
- completed development of a nationwide system of protected areas on public land, and waters that are representative of the major ecosystems in each biogeographical region;
- implemented management plans for protected areas identified by the Australian and New Zealand Environment and Conservation Council as having major conservation significance because of high biological diversity, high endemicity, or threatened species;
- established effective mechanisms for providing information to and support for biological diversity conservation projects undertaken by the community;
- clearly defined elements on the conservation of biological diversity in primary, secondary, and tertiary curricula, giving emphasis to inter-relationships between disciplines;
- implemented programs consistent with this Strategy designed to encourage local government to play a major role in nature conservation in Australia;
- implemented institutional arrangements and programs to ensure and monitor the ecologically sustainable development of Australian industries based on the extraction or

use of natural resources;
- arrested or reversed the decline of remnant native vegetation;
- avoided or limited any further broad-scale clearance of native vegetation, consistent with ecologically sustainable management and bioregional planning, in the instances in which regional biological diversity objectives are not compromised;
- completed species-specific management plans for major introduced pests and implemented effective controls for at least one introduced species of mammal and at least three major introduced plant pests;
- implemented a nationally coordinated program for long-term monitoring of the state of Australia's biological diversity and the impact of threatening processes;
- established legislative and administrative mechanisms for control of access to Australia's genetic resources;
- conducted an analysis of existing scientific knowledge about Australia's biological diversity and identified knowledge gaps and research priorities;
- fully implemented provisions of those international agreements relating to the conservation and sustainable use of biological diversity to which Australia is a signatory.

By the year 2005 Australia will have:
- established effective cooperative mechanisms for bioregional planning and management;
- implemented management plans for the protected area network;
- established a system of voluntary or cooperative reserves, or both, and other management schemes on private lands to complement the protection provided by the public estate in protected areas;
- established networks of community groups and volunteers that play major roles in managing and monitoring biological diversity at the district level;
- developed local governments that have assumed a major role in the conservation of Australia's biological diversity;
- demonstrated maintenance of regional and district floras and faunas;
- successfully rehabilitated at least 10 endangered or vulnerable species;
- successfully controlled three introduced mammals, 10 introduced plants, and controlled pathogens that pose major threats to biological diversity;
- developed sufficient information from long-term monitoring and other research to identify and understand the nature and extent of threats to Australia's biological diversity to develop actions for dealing with those threats.

Source: ANZECC, undated

ILLUSTRATIVE EXAMPLE 21

Step 4 Developing an Action Plan: Economic Incentives and Penalties that Promote Biodiversity Conservation

Economic incentives and penalties can be effective tools for enhancing implementation of the strategy and action plan and the conservation of biodiversity. The Indonesian Biodiversity Action Plan specifically proposed actions for developing economic incentives and penalties that promote biodiversity conservation.

ECONOMIC INCENTIVES AND PENALTIES THAT PROMOTE BIODIVERSITY CONSERVATION:

Economic incentives and disincentives should be built into all development programs and plans to exploit natural resources to restructure development policies for sustainable use.

Incorporate true environmental values and costs into project appraisals to ensure the sustainable use of resources and the environment.

Establish a "developer pays" policy requiring a developer to take measures to replant forests, prevent erosion, and mitigate other environmental damage caused by the development.

Require and evaluate environmental impact assessments for commercial operations and development programs which affect biological resources. Revoke licenses and/or implement strong financial disincentives on those who do not comply with measures to mitigate environmental damage.

Provide incentives to encourage *ex-situ* propagation programs for traded non-protected wildlife/plants that are easily bred in captivity to reduce the drain on wild populations.

Develop optimum management strategies to enhance economic returns from marketed goods and services in conservation areas in ways that complement rather than hinder ecosystem and biodiversity conservation efforts.

Source: GOI, 1993a

ILLUSTRATIVE EXAMPLE 22

Step 4 Developing an Action Plan: Bioregional Management

It is important to determine and clarify at which geographic scale the strategy or specific components of the strategy intends to address planning and management. The draft Australian Biodiversity Strategy addressed commitment to bioregional management by articulating an objective and proposing specific actions.

BIOREGIONAL PLANNING AND MANAGEMENT

OBJECTIVE

Manage biological diversity on a regional basis, using natural boundaries to facilitate the integration of conservation and production-oriented management.

Bioregional planning

Regional planning in which environmental characteristics are a principal determinant of boundaries is considered to be of major importance if biological diversity conservation is to succeed. The Murray-Darling Basin Commission, for example, plans on an environmental basis, using catchment boundaries as well as existing local, State, and Commonwealth structures. Several State and Territory governments are also beginning to plan and manage on a bioregional basis as part of their land management responsibilities. Actions such as this are needed elsewhere in Australia; they must be based on ecological parameters, vegetation types, catchment areas and climatic factors, combined with the interests of those living and working in the area.

One of the major determinants of the success of bioregional planning will be the extent to which all levels of government cooperate and coordinate their activities. For this cooperation to occur, a concerted nationwide effort is necessary to establish better lines of communication and coordination mechanisms that can be activated as soon as appropriate bioregional boundaries have been determined and accepted.

ACTIONS

Planning units

Determine principles for establishing bioregional planning units that emphasize regional environmental characteristics, are based on environmental parameters, and take account of productive uses and the identity and needs of human communities as appropriate.

These principles will include:

- identifying the biological diversity elements of national, regional, and local significance, the extent to which they need to be protected, and the extent to which they already occur in protected areas;
- identifying the major activities taking place within the region and in adjoining regions and analyzing how these may adversely affect the region's biological diversity, to ensure its use is ecologically sustainable;
- identifying any areas that are important for biological diversity conservation and require repair or rehabilitation;
- identifying priority areas for biological diversity conservation and for ecologically sustainable use, and their relationship to essential community requirements such as infrastructure and urban and industrial development;
- providing mechanisms for genuine, continuing community participation and proper assessment and monitoring processes;
- coordinating mechanisms to ensure ecologically sustainable use of biological diversity, with particular reference to agricultural lands, rangelands, water catchments and fisheries;
- incorporating flexibility, to allow for changes in land use allocation, including multiple and sequential uses of particular locations, and to accommodate improvements in knowledge and management techniques and changes in institutional arrangements.

BIOREGIONAL PLANS

Undertake bioregional planning for the conservation of biological diversity.

This planning will involve

- identifying appropriate intergovernmental and

Continued on page 76

Continued from page 75

intragovernmental mechanisms to ensure cooperation and coordination in bioregional planning;

- promoting the inclusion of biological diversity goals and principles in local government planning schemes and strategy plans;
- promoting sympathetic coordinated management of biological diversity for land and sea areas adjoining protected areas;
- improving protection of and management for biological diversity in closely settled environments and the coastal zone, with particular attention paid to corridors and remnants areas;
- increasing the number and involvement of those in the community who have special knowledge of biological diversity and skills in regional management, making use of existing community networks;
- providing suitably trained facilitators to help community participation, facilitate cooperation, and encourage resource managers to pursue ecological sustainability.

Source: ANZECC, undated

ILLUSTRATIVE EXAMPLE 23

Step 4 Developing an Action Plan: Access to Genetic Resources

Management of access to genetic resources is an essential provision of the Convention. The draft Australian Biodiversity Strategy formulated the country's objective concerning access to genetic resources and proposed a series of actions to meet their objective.

ACCESS TO GENETIC RESOURCES

OBJECTIVE

Ensure that the social and economic benefits of the use of genetic material and products derived from Australia's biological diversity accrue to Australia.

ACTIONS

Working group

Support the work of the ANZECC Task Force on Access to Australia's Genetic Resources and take prompt action to deal with its findings.

Sharing benefits

Through effective controls, legislation, and incentives (including secure property rights) ensure that Australia participates in research into and development of, and shares the benefits from, any commercial opportunities, including the development of biotechnologies that are based on genetic resources collected from areas within Australia's jurisdiction.

Non-threatening collection

Ensure that collection of genetic resources for research and development purposes does not adversely affect the viability or conservation status of the species or population being collected or of any component of its habitat.

Screening programs

Encourage and support the establishment of screening programs within Australia to identify genetic products of social and economic benefit.

Property rights

Ensure that Australia benefits from access to and use of its genetic resources through existing arrangements such as plant variety rights and patents legislation and any new arrangements that are developed.

The use and benefits of traditional knowledge are discussed under Action 1.8.2.

Ex-situ conservation of genetic material

Encourage and support the activities for Actions 1.9.1(b) and 1.9.1(c).

Source: ANZECC, undated

ILLUSTRATIVE EXAMPLE 24

Step 4 Developing an Action Plan: Project Formulation

Translating the strategy's objectives into practical priority actions is the essence of developing an action plan. The action plan should identify and articulate specific projects and include detailed information concerning its implementation. The draft Biodiversity Action Plan of Vietnam includes profiles of the priority actions it identified, such as the one that follows.

PROJECT CONCEPT

TITLE:	Development of Biodiversity Awareness Centre for Decision Makers
AIMS:	Increase the understanding of the importance of biodiversity conservation among senior government leaders and decision makers.
PRIORITY:	Medium
JUSTIFICATION:	Awareness among senior government leaders and decision makers about the importance of biodiversity conservation remains poor. The leaders are too busy to be retrained and have no time to read promotional literature. The only way to get their attention for a few hours is to take them away from their home and office environment for short but attractive learning weekends.
SCOPE:	• To establish three regional centres in attractive natural recreational sites (inside accessible parks or reserves) equipped to provide quick awareness and exposure to senior government officials. • To develop suitable awareness materials, audio visual programs, displays, courses and talks. • To hold periodic short seminars or weekends for senior leaders. The location and schedule must be attractive and recreational to attract busy officials from their normal duties for a few days. They will be temporarily spoiled but at the same time will absorb new ideas about the importance of biodiversity conservation. Courses would involve discussions led by skilled educators.
TIMING:	1996 for 5 years
LOCATION:	Ba Vi, Bac Ma, Nui Ba
RESPONSIBLE AGENCIES:	Ministry of Forestry, MOSTE
INTERNATIONAL PARTIES:	To be found
BUDGET:	U.S. $3 million
CBD ARTICLES:	6,13

Source: BAPPT, 1993

diversity being better conserved as the result of our work? Would some improvements have happened anyway? Are forestry, agriculture, and other uses becoming more sustainable? What else needs to be done to meet the goals? What lessons can be learned from this work? How can this new information and knowledge be made available to others? *(See Examples 25-27).*

STEP 7. REPORTING

A variety of written reports are sure to stem from the biodiversity-planning process. Here the distinction between the process and the product is drawn in more specific terms. In other words, the process of assessment, analysis, and formulation of measures for action and their implementation and monitoring (all within an environment of open dialogue with interested groups and stakeholders) may in itself already determine and launch the necessary action. The whole point of the process is to engender appropriate strategic action that will be of the most benefit to the country.

Well-written and well-presented documentation about the process and its impact on biodiversity will help achieve goals by providing information, widening the constituency, ensuring political compliance with mandates and commitments, and fostering partnership. That said, however, plans divorced from any activity may simply crowd the shelves of public offices and home libraries.

Three sub-steps are suggested:

First, determine which types of reports are needed, who the recipients are and should be, and which reporting interval is most appropriate.

For example, the program can prepare any or all of the following reports:

- annual status reports on the implementation of the national biodiversity plan, or its various sections, to the national chief executive, parliament, and the people;
- a country study assessing the nation's biotic wealth, challenges, and opportunities;
- a national strategy containing vision, goals, objectives, and measures for action;
- an action plan spelling out who will do what, when, where, how, and with what resources, along with measures for monitoring progress;
- a national biodiversity plan that summarizes the assessment, the gaps, the goals and objectives, priority measures, details on implementation, and measures for monitoring progress and evaluation;

BOX 15

Essential Elements to be Included in a Monitoring and Evaluation Program

The following indicators are among those that should be included in a monitoring and evaluation program:

- status and trends of the nation's use of terrestrial, aquatic, coastal, and marine resources, habitats, species, populations, genes, biodiversity services, and threats to biodiversity;
- shifts in selected social, political, and economic factors;
- shifts in human, institutional, facility, and funding capacity, including cultural practices and norms, technology, training and education, information availability, management, and monitoring capacity;
- changes in the policy and legal framework for natural resources, including protected areas, access to genetic resources, land tenure, property rights, benefit and cost sharing, trade and environmental impact assessment;
- changes in the use of biological resources and their sustainability, including natural-resource-based industries, and exploitation of resources for subsistence;
- trends in the monetary and non-monetary values of biodiversity and current expenditures and investments; and
- impacts of implementing the activities and policies of the biodiversity plan(s), vis-a-vis conservation, sustainability, and equity.

- five year status report on biodiversity and biological resources, showing rates and details of change in response to various factors, including the measures implemented in the biodiversity program;
- popular reports to the citizenry;
- early-warning bulletins that draw attention to species, genetic resources, and sites at risk, calling for appropriate action; and/or
- a periodic report to the Convention of the Parties, as called for in Article 26 (actions taken and results found during a particular reporting period, such as the cycle of the Conferences of the Parties), and a report to the Commission on Sustainable Development (actions taken to implement components of *Agenda 21*), among others.

Boxes 16 and 17 offer illustrative tables of contents for a national biodiversity strategy and an action plan. The structure and content of these documents provide policy makers and implementing partners with a brief walk through the process that the biodiversity-planning team followed to recommend actions, establish partnerships for implementation, agree to the schedule of implementation, and develop the budget. The documents should contain only that information required to support the decisions that need to be taken, to substantiate the recommendations, and to adequately inform users.

Obviously, should biodiversity planners choose to prepare a national biodiversity plan as a single

ILLUSTRATIVE EXAMPLE 25

Steps 5 and 6 Implementation and Monitoring: Objectives for Monitoring

The draft Canadian Biodiversity Strategy articulated specific objectives with which to measure the effectiveness of their strategy.

The Canadian Biodiversity Strategy will be a key building block in our efforts to achieve sustainable development. Ultimately, the degree to which the Strategy is able to enhance our national capacity to conserve biodiversity and achieve sustainable development will be the measure of its success. Specifically, we will know that the Strategy is making a difference if:

- the value and importance of biodiversity is reflected in the actions and decisions of all sectors of society from the large resource-based industries to private land-owners to the various orders of government;
- we are capturing existing information, generating new knowledge about biological resources and conveying that knowledge to decision-makers in a way that is useful, timely and efficient;
- we are no longer planning and making decisions based exclusively on a species-by-species and sector-by-sector basis, but are implementing an ecosystem approach to resource management and incorporating the full range of social, cultural, economic and ecological values;
- opportunities are being created through technological innovation, scientific discoveries and new applications of sustainable use; and
- we are maintaining Canada's biodiversity for future generations and contributing to conservation and sustainable use efforts worldwide through financial assistance, knowledge and expertise, access to our genetic resources, leadership, and by setting example.

Successful implementation of Canada's biodiversity strategy will require a coordinated approach based on cross-sectoral cooperation and partnerships among governments, non-government organizations, private sector interests, and individuals. The capacity to determine how Canada's biodiversity is managed is not limited to governments. Indigenous communities, businesses and industries, local communities and individuals must be involved with the implementation of the Canadian Biodiversity Strategy.

Source: FPT Biodiversity Working Group, 1994

ILLUSTRATIVE EXAMPLE 26

Step 6 Monitoring: Components of a Monitoring System

The China Biodiversity Action Plan includes a box focusing on the importance of information and a sound monitoring system for conservation activities.

OBJECTIVES AND COMPONENTS OF A BIODIVERSITY MONITORING SYSTEM FOR CHINA

Sound decisions on the conservation of biological diversity are based on detailed and accurate information. For instance, knowledge of land cover conditions and land use changes is an important prerequisite to conservation plans. An effective monitoring system to provide this kind of detailed information about the environment and the results of policies and actions, would have the following features.

OBJECTIVE 1

Present the results of monitoring in a form readily available and understandable to scientists, managers, and the public.

- Use GIS to analyze present biogeographic information and to aid monitoring.
- Publish results of monitoring activities on a timely basis.

OBJECTIVE 2

Provide information on changes in land classification, use, and ecosystem health.

- Prepare and maintain a national biogeographic (ecosystem) database.
- Establish a continuous remote sensing system for observing and monitoring climate and land use changes, plant disease outbreaks, and other environmental problems as technology permits.
- Establish monitoring stations for water flows and quality in critical watersheds.

OBJECTIVE 3

Provide accurate and timely information on population size and trends, especially of threatened species.

- Include ecological information as part of forest inventories.
- Periodically conduct surveys of threatened species of birds and other animals.
- Determine if indicator species can be used to monitor ecosystem changes.

OBJECTIVE 4

Provide information on the effects of airborne pollutants on biodiversity.

- Complete air monitoring station network with emphasis on NO_X, SO_X, particulates, CO_2 and CO.
- Establish acid deposition monitoring stations, analyze information and study effects.
- Study the direct effects of airborne toxicants on vegetation and soil organisms.

OBJECTIVE 5

Monitor implementation of policies and projects.

- Examine national and provincial budgets to determine if resources are being properly allocated, so to carry out the biodiversity action plan.
- Existing legislation should be reviewed to determine consistency with this plan.
- Inspect and carefully examine ongoing and completed projects to evaluate success.
- Establish regular foot patrols in reserves; train forest guards to monitor conditions in and outside of reserves.

Source: NEPA, 1994

ILLUSTRATIVE EXAMPLE 27

Step 6 Monitoring: Linkages between Data, Information, Monitoring and Development of a Biodiversity Strategy and Action Plan

Information collected throughout the assessment and strategy phases should be used actively to inform decisions on conservation and management, and to provide a scientific basis for actions. The Biodiversity Action Plan of the United Kingdom graphically represented the linkages between data, information, the strategy process, and the Convention.

DATA, INFORMATION, AND MONITORING UNDER THE CONVENTION ON BIOLOGICAL DIVERSITY

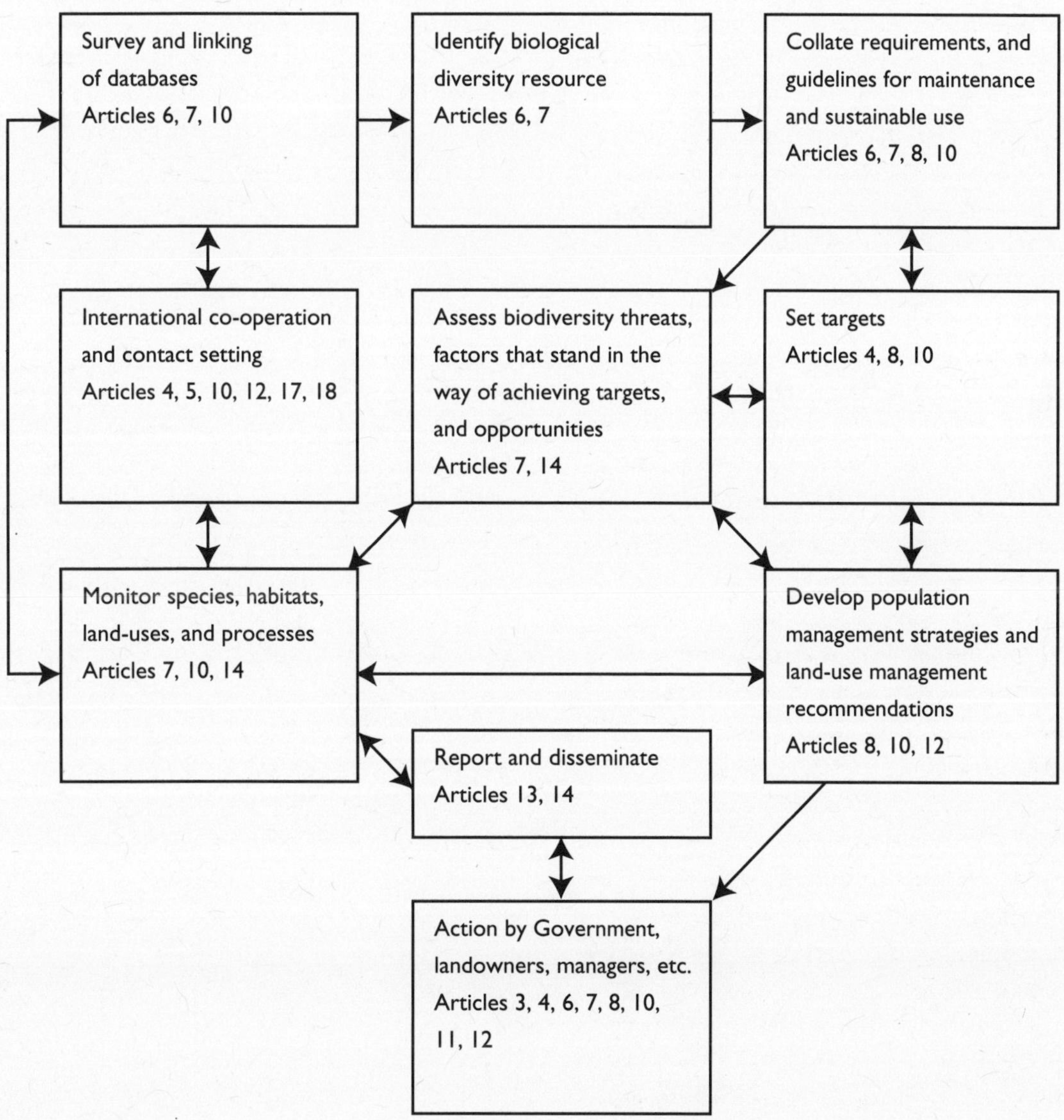

Note: Requirements for data collection, monitoring and other activities under the Convention. Each action interlinks with others to form the data and information base for Action Plan.

Source: Department of Environment (UK), 1994a

BOX 16

Illustrative Table of Contents for a National Biodiversity Strategy Report

1. EXECUTIVE SUMMARY

Briefly summarize the strategy report, stating succinctly the importance of biodiversity, the commitment to the Convention, the mandate, the participants' list, the biotic wealth and national capacity, the goals and gaps, and strategic recommendations.

2. INTRODUCTION

Describe why biodiversity is important to the country and its local communities. Explain the Convention and the nation's commitment to its provisions. Present the aim of the national strategy and describe whom it is trying to reach.

3. BACKGROUND

Describe the legal and policy framework that provides the mandate and instructions for preparing the strategy report. Explain the ongoing biodiversity work in the country and how it is organized. Who is involved in planning and implementation? How and when are they being involved? What is the process of dialogue and negotiation? Describe the stakeholders and their legitimate interests in biodiversity and its allocation and management. Explain why their participation is key to successful implementation of the plan and long-term survival of the nation's biotic wealth.

4. ASSESSMENT

Summarize the nation's biotic wealth, capacity (personal, institutional, facility, and financial), and ongoing programs. Describe and analyze the key factors driving biodiversity loss (why they exist, what effect they have, what constituencies they serve, what changes they make necessary, and what it will take to make these changes). Refer through graphics to the status and trends in the nation's wild plants, animals, microbes, habitats, landscapes and waterscapes; data-management capability and available services; personnel capability, training, and education in related fields; institutions, including cultural practices and norms, organizations and their mandates and capabilities, land tenure; major uses of biological resources, current levels of sustainability; status of the nation's technology development and use; the monetary and non-monetary values involved (who receives benefits and pays the costs); financial flows from internal and external sources; and gaps in information.

5. GOALS AND OBJECTIVES

Create a vision for biodiversity and its place in society, its protection, understanding, sustainable use, and the equitable sharing of its benefits and costs. Specify targets for meeting national, local, and international goals in terms of protecting, assessing, using, and benefiting from biodiversity and its components.

6. GAPS

Describe and characterize the gaps between the current reality and the vision: summarize the current status and trends in habitats and ecosystems, species, and special genetic traits or strains that warrant attention; the factors driving biodiversity loss; the capacity of the country in terms of the skills of those who work on biodiversity management, institutions, facilities, financial mechanisms, and technology. Compare this capacity to the aspirations and vision of the country as stated in the goals and objectives. List the gaps.

7. STRATEGIC RECOMMENDATIONS

State the activities, policies, and tasks that have been selected for implementation that promise to cover the gaps. Classify them according to the goals and objectives and the articles to the Convention. Give the priorities for each activity, policy, and task. Mention any particularly noteworthy aspects of the negotiation process by which these choices were made by the involved stakeholders.

BOX 17

Illustrative Table of Contents for a National Biodiversity Action Plan Report

1. EXECUTIVE SUMMARY

Briefly summarize the action plan report, stating succinctly the importance of biodiversity, the commitment to the Convention, the mandate, the participants' list, the biotic wealth and national capacity, the goals and gaps, strategic recommendations, and characteristics of the action (who will do what, when, where, with what means and funding).

2. INTRODUCTION

Describe why biodiversity is important to the country and its local communities. Explain the Convention and the nation's commitment to its provisions. Present the aim of the national biodiversity action plan and specify to whom it is directed.

3. BACKGROUND

Describe the legal and policy framework that provides the mandate and instructions for preparing the action plan report. Provide a short summary of the nation's biotic assets, capacity (personal, institutions, facilities, and funding), and ongoing programs. Explain the institutional arrangements and responsibilities so people will know how the strategic recommendations will be implemented.

4. GOALS AND OBJECTIVES

State the vision for biodiversity and its place in society, focusing on its protection, on scientific understanding, on sustainable use, and on the equitable sharing of its benefits and costs. Specify targets to meet the national, local, and international goals in terms of protecting, assessing, utilizing, and benefiting from biodiversity and its components.

5. THE STRATEGY

Summarize the gaps between the current situation in the country and the stated vision, goals, and objectives. Summarize the strategic recommendations, including the activities, policies, and tasks that have been selected for implementation that promise to cover the gaps. Give priorities for each.

6. THE PARTNERS

Describe the public and private entities, communities, and industries that have participated in the process and have agreed to take responsibility for particular activities and investments.

7. THE ACTION

Present the detailed activities, tasks, and policies to be implemented. Explain which partner (ministry, industry, indigenous group, NGO, university) will implement each item, where, and what measures the partners will employ.

8. SCHEDULE

Present a timetable for implementation of the various tasks, reflecting the priorities that have been assigned. Note mileposts to help signal progress or delay.

9. BUDGET

Provide the budget for the plan of action, showing funding requirements for operating expenses, capital purchases, transport, field costs, etc. List the personnel needed by category of skill or background, the facilities and services required, and possible international technical and financial cooperation.

10. MONITORING AND EVALUATION

Explain the measures to be used for tracking the results of the action plan and for monitoring changes in the economy, environment, and society. Give the indicators that will be used. Present the individuals and organizations who will carry these responsibilities and how they were selected. Note the audience for the reports, along with the document's content and timing of implementation.

comprehensive document (to combine the strategy and action plan into one report) the two illustrative outlines in Boxes 16 and 17 could easily be merged and integrated.

Second, draft the reports, adequately balancing content and information.

These documents can easily get bogged down in a thicket of descriptive material on resources and institutions, losing sight of the primary aim—to provide guidance for policy decisions, inform key constituencies, and foster wider awareness and commitment. The content of early draft documents should thus be edited to balance three fundamental dimensions in the presentation: (a) descriptive information about resources and institutions, legislation, policies, and programs; (b) analysis of issues and options, gaps, and opportunities; and (c) prescriptions for action. The tendency is to provide excessively detailed descriptive information on resources and institutions, limited analysis of issues and options, and rough sketches of the actions needed.

Third and finally, promote the findings and proposals contained in the reports.

The impact of these plans can be greatly enhanced by ensuring their adequate distribution and presentation to selected target audiences. An "outreach strategy" can include such components as:

- Public meetings that provide information about biodiversity and its values to people; the distribution of free copies of the strategy and action plans to interested groups and individuals; and such other promotional mechanisms as media presentations, exhibits, and special events. Enough copies of the strategy and action plan should be published in local languages and a relevant international language so that it can be distributed widely among key constituents at the local, national, and international levels. During this phase of the work, artists and performers can be called on to build the concepts of the strategy into their public work.
- Copies of the strategy and action plan should be sent to the news media with kits that provide materials in formats and languages that communicate well to the general public. Well-known national figures and media personalities can be recruited to hold press conferences at the launch of the strategy and action plans into the public arena.
- Featuring and explaining the strategy at conferences, workshops, and community meetings. Care should be taken to use the most culturally appropriate communication tools for each constituent group.
- Presenting the strategy at international meetings, especially those sponsored by United Nations organizations and UN specialized agencies, international NGOs, bilateral and multilateral donor and financial institutions, and other international partner organizations.

Annex A • Profiles in National Biodiversity Planning

Annex A: Profiles in National Biodiversity Planning

In the short period since the United Nations Environment Programme launched its country study process in 1991, countries around the world have prepared national biodiversity plans. They have included biodiversity strategies, action plans, country studies, and biodiversity programs. This annex was prepared to provide the reader with succinct summaries of those countries' experiences. These profiles were drawn from detailed case study reports prepared by key individuals from collaborating countries and are grouped by the type of planning effort.

■ 1. *The Australian National Biodiversity Strategy*[1]

INSTITUTIONAL BASE

The Federal Department of the Environment, Sport and Territories (DEST)—specifically, the Biodiversity Unit within the Environment Strategies Directorate—initiated the preparation of the National Strategy for the Conservation of Australia's Biological Diversity. The mandate came from the former prime minister of Australia in 1989 and was reaffirmed by the current prime minister.

METHODOLOGY

The Biodiversity Unit developed a six-step methodology for preparing the national strategy.

1. Biological Diversity Advisory Committee

The Minister responsible for the environment convened an 11-member Biological Diversity Advisory Committee (BDAC) in 1991, initially representing wide-ranging scientific disciplines, non-governmental conservation organizations, two state governments, and the National Farmers Federation. Later, four members were added from the forestry, fishing, tourism, and mining industries. The Committee met 11 times between April 1991 and August 1992. It analyzed the status of and threats to Australia's biodiversity, the adequacy of existing mechanisms, and the relevant international developments. Then it prepared the first draft of the strategy.

2. Public consultation process

The draft strategy was available for public comment from March to May 1992.

3. National conference

A national conference was sponsored by DEST and the Ecological Society of Australia to encourage debate on the national strategy. At 10 conference workshops, chapters of the draft strategy and key issues relevant to the strategy were discussed.

4. *Presentation to minister and referral to ANZECC*

BDAC presented a draft strategy to the minister responsible for the environment in September 1992. The minister referred the draft strategy to the Australian and New Zealand Environment and Conservation Council (ANZECC), composed of environment ministers from federal, state, and territory governments.

5. *Final drafting by ANZECC Task Force*

A task force under ANZECC considered the further development and finalization of the National Strategy.

6. *Government-wide consideration*

In October 1993, the ANZECC Task Force on Biological Diversity submitted a final draft of the strategy to ANZECC, whose members sought consideration of the strategy by the Australian government (in each of the States and Territories and the Commonwealth).

PARTICIPATION

Many of the interest groups and institutions invited to help prepare the national strategy participated fully in the Advisory Committee and the ANZECC Task Force. An even broader range of participants took part in bilateral or multi-lateral meetings and public consultation, including all relevant ministries and other members of Parliament; government departments and agencies at all levels; scientific, research, and academic institutions; business and industry; nongovernmental organizations; professional societies; education institutions; advisory councils; and interested individuals.

Membership in the ANZECC Task Force comprised representatives from all ANZECC agencies and representatives of the Agriculture and Resources Management Council of Australia and New Zealand, the Australian Forestry Council, the Australian and New Zealand Fisheries and Aquaculture Council, the Australian and New Zealand Minerals and Energy Council, and the Industry, Technology and Regional Development Council. The Task Force met six times and considered fully all government interests in finalizing the strategy.

GOALS AND OBJECTIVES

The overall goal of the national strategy is to protect biological diversity and maintain ecological processes and systems. The seven operational goals are: (1) conserving biological diversity across Australia; (2) integrating biological diversity conservation and natural resource management; (3) managing biodiversity-threatening processes; (4) improving knowledge; (5) involving communities; (6) defining Australia's international role; and (7) setting priorities. Aligned under these seven goals are 35 major objectives.

RELATIONSHIP TO NATIONAL DEVELOPMENT PLANNING

Australia's national biodiversity strategy was developed in close accordance with the National Strategy for Ecologically Sustainable Development (NSESD), which outlines Australia's national approach to development. The strategy acknowledges other relevant national processes that also contribute to ecologically sustainable development, including the National Forest Policy Statement, the National Greenhouse Response Strategy, the draft National Rangelands Strategy, and the National Strategy for the Conservation of Australian Species and Ecological Communities Threatened with Extinction. Each recognizes the need to conserve biological diversity, and the national biodiversity strategy indicates inter-linkages with them in relation to specific objectives and actions.

INTENDED TARGET OF PLANNING EFFORTS

The reason for developing the strategy is to provide information and policy guidance for decision-makers. The strategy outlines actions that all relevant government agencies should take in the areas of environment and conservation, resource and industrial development, research, education, and other areas. It also recognizes the roles of industry, business, and nongovernmental organizations in biodiversity conservation. Specific issues and actions for various sectors of the community are also identified.

RELATIONSHIP TO THE CONVENTION ON BIODIVERSITY

The national strategy explicitly takes account of the provisions of the Convention on Biological Diversity. The Biodiversity Unit of DEST had played a major role in the negotiations for the Convention and led efforts to draft and redraft the national strategy. Provisions of the Convention were compared and analyzed with a view to incorporating all relevant provisions into the national strategy and to ensure consistency. In many respects, the national strategy reflects the general structure of the Convention.

SCALES OF PLANNING EFFORTS

To ensure integrated planning and management approaches, the national strategy is organized mainly around two spatial scales: continental and bioregional. The local scale is also recognized in the strategy, though activity at this level has been subordinated and integrated into the bioregional planning framework. The strategy recognizes biodiversity shared with neighboring countries, though no other countries were formally involved in its preparation.

INTERNATIONAL ASSISTANCE

Australia's biodiversity-planning effort received no technical or financial support from international sources.

MONITORING AND FOLLOW-UP

The national strategy outlines the arrangements required for implementing the plan and for monitoring its effectiveness. The overall responsibility for implementation is assigned to ANZECC, which is to consult with other relevant ministerial councils. ANZECC is to monitor the outcomes of the strategy, report to heads of government on its implementation, and provide five-yearly reviews of its progress. The Biological Diversity Advisory Council manages the input on implementation of the strategy from scientists, nongovernmental organizations, and industry.

OBSTACLES

The development of the national strategy encountered several barriers:

- lack of understanding of the concept of biodiversity conservation
- difficulty in operationalizing the concept
- need for further research on the role and function of biodiversity in ecological processes
- lack of appropriate methods of biodiversity management
- need for improved communication between scientists and policy-makers
- diversity of philosophical positions and views
- poor mechanisms to fully value biological resources and ecological functions.

FACILITATING FACTORS

The development of the national strategy received political commitment at the highest level and enjoyed the continued interest and support of the previous and current federal ministers for the environment. Preparation in parallel with negotiations of the Convention on Biological Diversity also provided another dimension to arguments for a strategy for Australia. Domestically, association of the strategy with the overarching NSESD also gave the strategy impetus. The support of nongovernmental conservation organizations helped keep the process moving while the cooperation and involvement of business and industry increased the strategy's credibility. Overall, broad community consultation ensured that "ownership" of the final product was not just restricted to government and interest groups.

Other facilitating factors included a strong scientific rationale for biodiversity conservation; a well established base of existing legislation, policies, and programs, including those that integrated conservation with sustainable use; and a range of concurrent research and communication activities.

COST, PERSONNEL, AND TIME REQUIRED

The development of the National Strategy for the Conservation of Australia's Biological Diversity has been a complex process. Because a cooperative approach between federal and state and territorial governments was required, along with consultation and involvement of a broad range of stakeholders, the development process, now in its final stages, has taken three and a half years.

At the federal level, funds and staffing resources were dedicated through the budget process, which included monies to support increasing public awareness about biodiversity conservation issues.

1. Based on case study material prepared by Josephine Mummery and other members of the Biodiversity Unit (including Tas Sakellans, Andreas Glanznig, David Boughey, and Rod Holesgrove [Director]), Department of the Environment, Sport and Territories, Canberra, Australia.

2. *The Canadian National Biodiversity Strategy*[2]

INSTITUTIONAL BASE

Preparation of a Canadian Biodiversity Strategy is being led by a working group consisting of representatives from each of the federal, provincial, and territorial governments. The activity of the Federal/Provincial/Territorial Biodiversity Working Group (F/P/T Working Group) is being coordinated by the Biodiversity Convention Office (BCO).

Although the BCO is based in Environment Canada, a federal government department, it functions as a national coordinating body. It coordinates input from all relevant federal government agencies and represents the interests of the F/P/T Working Group.

The BCO's current mandate was established in November 1992. Federal, provincial, and territorial ministers of environment, parks, forestry, and wildlife endorsed a "follow-up plan" for the Convention that charged the F/P/T Working Group with developing a biodiversity strategy and the BCO with convening and serving the Working Group.

METHODOLOGY

The process for preparing a biodiversity strategy was developed by the F/P/T Working Group in consultation with its nongovernmental advisory group. Development of the strategy can be broken down into five distinct steps:

1. Inventory

Current policies, programs, activities, and legislation relevant to biodiversity and the articles of the Convention were surveyed.

2. Analysis

Each participating organization and jurisdiction gave the BCO a list of proposed recommendations, based on the Inventory, to take advantage of identified opportunities or to fill identified gaps and weaknesses in policies, programs, and legislation seeking to meet Convention objectives.

3. Drafting

The BCO coordinated the drafting and internal review of the strategy through consensus-building workshops.

4. Stakeholder review

This consultation effort will focus primarily on interested stakeholders, but the general public can also comment.

5. Cabinet approval

Once the advice of stakeholders has been received and considered, the strategy will be finalized and submitted for cabinet approval at each level of government.

The Canadian Museum of Nature is coordinating a Canada Country Study. Currently in final draft form, the material in the Country Study is being used in the development of the Biodiversity Strategy.

PARTICIPATION

1. The Federal/Provincial/Territorial Working Group

This decision-making body is composed of one representative from each province and territory and from key federal departments, including Fisheries and Oceans, Natural Resources, and Agriculture.

2. The Biodiversity Convention Advisory Group (BCAG)

Originally established to provide input and advice on the negotiation of the Convention, BCAG has participated in all stages of developing the strategy since UNCED in 1992. It is composed of a broad range of representatives from indigenous groups, industry, conservation groups, academia, and the scientific community.

3. Expert Focus Groups

Ten expert focus groups convened to discuss certain subjects on biodiversity and to develop strategic recommendations.

GOALS AND OBJECTIVES

The proposed goals of the draft Biodiversity Strategy are:

- to conserve biodiversity and sustainably use biological resources
- to develop and implement ecological management approaches to conserve and sustainably use biodiversity
- to promote an understanding of the need to conserve and sustainably use biodiversity

- to maintain or develop incentives and legislation that supports the conservation of biodiversity and the sustainable use of biological resources, and
- to work with other countries to conserve biodiversity, use biological resources sustainably, and share equitably the benefits of using genetic resources.

The Biodiversity Strategy also supports the development of contributions from indigenous people to the strategy-planning exercise.

RELATIONSHIP TO NATIONAL DEVELOPMENT PLANNING

Although Canada has no official national development plan, the Biodiversity Strategy builds numerous federal, provincial, and territorial policies, plans, and strategies. In particular, an underlying theme of the Biodiversity Strategy is improved coordination and increased harmonization of policies, programs, and legislation to reduce overlap and eliminate duplication of efforts and activities, and to create a cohesive policy framework for conserving biodiversity and sustainably using biological resources.

INTENDED TARGET OF PLANNING EFFORTS

The purpose of the draft Biodiversity Strategy is to provide policy guidance for those making decisions about biodiversity in Canada. It is aimed primarily at governments, but the process adopted was designed to develop a broad base of support from important national organizations (such as those represented on the BCAG) and to generate cooperation among communities and individuals.

RELATIONSHIP TO THE CONVENTION ON BIODIVERSITY

The planning exercise explicitly addressed the articles of the Convention. The Inventory and Analysis phases of work tracked the Convention article by article. The draft strategy explicitly references and quotes Convention articles in every chapter as a strategy base.

SCALES OF PLANNING EFFORTS

The biodiversity strategy is national in scope, although the planning exercise focuses on both national and international issues. Provincial, regional, and local issues will be addressed more fully during action planning and implementation, when the strategy goes into effect at all levels of government. Implementation will involve cooperation and partnerships among governments, nongovernmental organizations, the private sector, communities, and individuals. The proposed goals and commitments of the strategy take into consideration the shared responsibility for conserving Canada's biodiversity, as well as international conservation agreements.

No neighboring countries have been directly involved in the development of the Biodiversity Strategy, though many cooperative biodiversity-related programs exist, particularly with the United States.

INTERNATIONAL ASSISTANCE

Canada received no technical or financial support from international sources for the development of the Biodiversity Strategy.

MONITORING AND FOLLOW-UP

Monitoring, reporting, and evaluating processes will be important to the strategy but have not been finalized.

OBSTACLES AND CHALLENGES

1. Political

Integrating the views and values of Canadian citizens presented government officials with many challenges, both in managing the process and in addressing the concerns of very different stakeholders, including business, academia, and conservation groups. The evolving political status of indigenous peoples in Canada and the resulting shifts in political sovereignty pose special challenges during public policy development.

2. Constitutional

Under the Canadian constitution, federal, provincial, and territorial governments share the responsibility for managing biological resources and for economic development. This results in complex negotiations among 13 jurisdictions.

3. Institutional

Responsibility for the conservation and sustainable use of biodiversity is not centralized in any one government federally, provincially or, territorially. Communication and coordination on the development of the strategy are, therefore, not only inter-governmental issues, but also intra-governmental issues at the federal, provincial, and territorial levels.

4. Financial

The expense of ensuring the development and maintenance of a consensus among the federal, provincial, and territorial governments on a plan for biodiversity has been significant in terms of the financial and human resources required, and it has limited participation of some governments, NGOs, and community members.

FACILITATING FACTORS

1. Existing legislation, policies and programs

Canada has a well-established policy base on which to build a national biodiversity strategy. At the federal, provincial and territorial levels, numerous conservation and sustainable development initiatives are in place or being developed.

2. Policy development experience

Canada has extensive experience with stakeholder consultation and participation in public policy development.

3. UNCED preparation

Multi-stakeholder advisory groups established for the negotiations of the Biodiversity Convention and preparations for UNCED to provide advice from the private sector, academia, and conservation groups have proven to be valuable to the strategy-development process.

4. Existing biodiversity activity

Many other non-Convention-related initiatives on biodiversity in Canada have provided opportunities for effective and efficient sharing of expenses, expertise, experience, knowledge, collaborative projects, etc.

COST, PERSONNEL, AND TIME REQUIREMENTS

The federal government has allocated approximately C$1.4 million for developing the strategy between January 1993 and November 1994. The Canadian Council of Ministers of the Environment has allocated $225,000 to the task.

Approximately 300 people are working on the Biodiversity Strategy, though only Biodiversity Convention Office staff are working on the Biodiversity Strategy full-time. This includes 9 staff members from the Biodiversity Convention Office (secretariat), 18 from the Federal-Provincial-Territorial Working Group, 28 from the Biodiversity Convention Advisory Group, 45 from the Federal Interdepartmental Committee on Biodiversity, and 200 from the 10 Expert Focus Groups.

STATUS OF PLAN

A draft for discussion was released in June 1994.

2. Based on case study material provided by John Herity, Director, Biodiversity Convention Office, Environment Canada, Quebec, Canada.

3. The Philippine National Biodiversity Strategy[3]

Faced with some of the greatest loss of biodiversity in the world, the Philippines has initiated various efforts to conserve the nation's biodiversity over the past five to 10 years. Yet, these efforts more closely resemble rescue missions for the remaining natural habitats than comprehensive biodiversity planning. More comprehensive efforts to develop a national action plan for biodiversity conservation did, however, begin in 1988—the same year that the new cornerstone of national biodiversity conservation planning and action, the Integrated Protected Areas System (IPAS), was established in the Philippines. In 1987, the Department of Environment and Natural Resources (DENR) initiated a process to formulate a Philippine Strategy for Sustainable Development (PSSD) with sections on forests, protected areas and biodiversity, freshwater ecosystems, and coastal resources.

In 1992, in the wake of the UN Conference on Environment and Development (UNCED), the Philippine Council for Sustainable Development (PCSD) was created as a quasi-governmental body by Presidential Order. The director general of the

National Economic and Development Authority is chair, the secretary of DENR is vice-chairman, and 14 representatives of government departments and seven NGO representatives are members.

In April 1994, the final Philippine Strategy for Biological Diversity Conservation was approved by the president.

The Protected Areas and Wildlife Bureau (PAWB) of DENR plans to undertake a biodiversity country study under the auspices of UNEP.. This effort is expected to provide a more solid information base for developing the work plans under the strategy's 18 objectives, for prioritizing those objectives, and for assisting in identifying lead or responsible government agencies.

INSTITUTIONAL BASE

The DENR, particularly its Protected Areas and Wildlife Bureau, is the lead agency on biodiversity conservation. The establishment of the PCSD and its Subcommittee on Biodiversity has strengthened the hand of PAWB (whose director is also chair of the Subcommittee) and given it a channel for systematic dialogue with sectoral agencies and NGOs. NGOs have been in the vanguard of raising biodiversity as an issue and, with the creation of the Foundation for the Philippine Environment (FPE) and NGOs for Integrated Protected Areas, Inc. (NIPA), they became important players in designing and implementing biodiversity conservation programs and projects.

METHODOLOGY AND PARTICIPATION

Broad parameters and priorities are set at workshops, where ongoing activities are inventoried. Then, a draft plan is prepared by the lead government agency. Consultative meetings and revisions of the draft follow, leading to the finalization and legitimation of the plan through the PCSD, DENR, the cabinet, and the president.

RELATIONSHIP TO NATIONAL DEVELOPMENT PLANNING

It is too early to say what impact the strategy has had on national development planning because the newly completed strategy is a framework awaiting particulars on sectoral activities and policy directions, among other things. However, the resolution creating the PSSD notes, "The translation of the PSSD conceptual framework will have reached its final cycle with the preparation of a Government Action Program," which, "while specifically addressing the concerns of a strategy for sustainable development will be treated as an integral component of the country's national development plan." If the Subcommittee's work is accepted by the PCSD chair (the National Economic Development Authority, NEDA), it is likely to be integrated into the country's mainstream development planning.

INTENDED TARGET OF PLANNING EFFORTS

As noted above, the biodiversity-planning process used several multi-sectoral meetings and workshops to gather and disseminate information, and to involve a wider circle of decision-makers, experts, and activists in the process. Its integration into the PCSD structure also means that the strategy will reach a wider group of decision makers than would be the case if, for example, the DENR had prepared the document alone.

RELATIONSHIP TO THE CONVENTION ON BIOLOGICAL DIVERSITY

The Philippines was an active participant in the negotiation of the Convention and ratified it in October 1993. This Convention process and participation in the UN Conference on Environment and Development have catalyzed national initiatives. The national Strategy for Biological Diversity Conservation notes that it is "the Philippines' contribution to the implementation of *Agenda 21,* the Convention on Biological Diversity, the Philippine National Integrated Protected Areas System Act (NIPAS), and the philosophy being promoted in *Caring for the Earth.*" On the other hand, biodiversity planning in the Philippines began in the mid-1980s with the development of the IPAS system, NIPAS Act law, and the Philippine Strategy for Sustainable Development. In short, the strategy contributes to the Convention and responds to many of its provisions, but it was not carried out because of the Convention.

SCALES OF PLANNING EFFORTS

As important as work at the bioregional level under the IPAS program is, provincial-level planning exercises will most likely be necessary—especially in the context of the ongoing decentralization under the Local Government Code—and there has been some discussion about launching such efforts in the coming year or so. Meanwhile, some resource-specific planning has taken place, including finalization of the *National Wetland Action Plan* produced in late 1992 by DENR in cooperation with the Asian Wetlands Bureau.

INTERNATIONAL ASSISTANCE

Biodiversity planning in the Philippines has been predominantly driven by the Philippines' own agenda. Yet, donors have provided financial support to both DENR and to local and national NGOs.

The World Wildlife Fund-US has been an influential donor since at least 1986. It has provided financial and technical assistance, including grants, a debt-for-nature swap program, natural-resource-management planning, and a biodiversity training center in the Subic Forest.

The Global Environment Facility, through the World Bank, is providing US$20 million in grant funds to implement IPAS management at 10 key sites. The U.S. Agency for International Development is funding a large ($125 million) Natural Resources Management Project (NRMP), with significant biodiversity components.

Several bilateral funding agencies and various private foundations (e.g., the MacArthur, Ford, and Asia Foundations) support biodiversity conservation activities but have not been directly funding the Philippine planning process *per se*.

MONITORING AND FOLLOW-UP

The strategy will be further developed as sub-strategies and workplans for the 18 objectives the strategy lays out. The PCSD, through its Biodiversity Subcommittee, will presumably play a lead role in monitoring follow-up activities. But NGOs—especially the many that are part of NIPA, Inc., and the FPE—are also likely to play a large role, both as members of the PCSD and in their independent activities.

OBSTACLES

Sectoral divisions and competition have hampered the biodiversity-planning process. The Philippines' generally weak economic situation has also been a factor because any significant increases in biodiversity funding to meet planning goals must come from external donors.

Although domestic and transnational corporations are deeply implicated in the country's rapid loss of biodiversity, the planning process has neither confronted these strong vested interests nor explored ways to tap their resources. One exception is the involvement of Philippine Business for Social Progress—an association of Philippine businesses with at least a rhetorical commitment to sustainable development—in some biodiversity activities, such as the 1988 debt-for-nature swap.

FACILITATING FACTORS

In the Philippines' political system, biodiversity planning has drawn on a far broader range of interests and groups from the civil society than similar processes in neighboring countries have. The relative strength of the NGO community, which has provided many perspectives and much data, has also been a positive factor, as has the country's comparatively strong community of academic biodiversity specialists.

Preparations for the Earth Summit and the resulting establishment of the PCSD have catalyzed action in the Philippines, as has the biodiversity convention process. Finally, the truly critical situation of the country's biodiversity has provided a sense of urgency.

RESULTS TO DATE

Of the many biodiversity conservation activities that have taken place in the past five years, some—such as the IPAS program—have already resulted in action in the field. The full results of the new national biodiversity strategy, however, will not be known for some time to come.

3. Based on case study material provided by Charles Barber, Senior Associate, World Resources Institute, through interviews with Jose Ampeso (Philippine Department of Foreign Affairs), Mary Jean Caleda (Philippine Department of Environment and Natural Resources), Corazon Catibog-Sinha (Philippine Department of Environment and Natural Resources), Richard Edwards (WWF-Philippines), Delfin Ganapin (Foundation for the Philippine Environment), Christie Nozawa (Haribon Foundation), and Rene Salazar (SEARICE).

■ 4. *The German National Biodiversity Strategy*[4]

INSTITUTIONAL BASE

The German Federal Ministry for the Environment Division for General Questions of Nature Conservation is responsible for launching the national strategy to implement the requirements of the Convention on Biological Diversity.

The actual power concerning conservation issues lies within the federal states (the "Laender"). The Laender must fulfill the Federal Nature Conservation Act of 1976 with their own State Conservation Laws. For international agreements, the federal government must oversee the Laender to ensure that they fulfill the obligations under the Convention.

METHODOLOGY

The methodology for the strategy is framed by the objectives of several documents: the Convention on Biological Diversity, *Agenda 21* (in particular, Chapters 14 and 15), the *Global Biodiversity Strategy,* other international and multilateral nature conservation agreements, European policies and regulations, and the German UNEP Country Study. The methodology is also informed by a participatory process involving Parliament, Laender, NGOs, and the private sector.

The Ministry for the Environment used four guiding principles in the process:

- balancing timely development of a strategy with the time needed for participation and review by interested parties
- emphasizing participatory planning as central to ensure that the strategy is planned and implemented efficiently
- defining the scope of the strategy process to integrate biodiversity considerations into the overall environmental planning process and into decision making at all levels of government (target group)
- establishing a steering group to launch, orient, and facilitate the strategy process.

The Ministry drafted a strategy paper to provide relevant and reliable information to decision and policy makers at key opportunities. The final strategy paper will be disseminated to members of Parliament, the federal ministries, the Laender governments (which will disseminate it to district- and county-level officials), NGOs and the interested public.

PARTICIPATION

The following groups and institutions were either invited or had the constitutional right to participate and review:

1. Government

All ministries were entitled to be involved. The following ministries participated: Foreign Affairs, Economy, Finance, Economic Cooperation and Development, Health, Food and Agriculture, Justice, Construction, and Land Use Planning.

2. Parliament

The Parliament (Bundestag) approved the ratification of the Convention in June-July 1993.

3. Laender

The Laender approved the ratification of the Convention in June 1993. The actual implementing power concerning conservation issues lies solely within the Laender.

4. Nongovernmental Organizations (NGOs)

Interested groups (in particular NGOs in conservation and development) participated in the process from the start. The NGO German Biodiversity Working Group serves as the contact point between government and NGOs.

5. Private Sector

While the private sector is welcome to participate in the strategy process, there has been very little involvement.

GOALS AND OBJECTIVES

The objectives of the strategy reflect the goals of the Federal Nature Conservation Act: conservation, preservation, and development of nature and landscapes, in both populated and unpopulated areas, to serve the following purposes:

- to maintain the efficiency of the balance of nature
- to preserve the exploitability of nature's resources
- to conserve fauna and flora, and
- to safeguard the variety, particularity, and beauty of nature and landscapes.

RELATIONSHIP TO NATIONAL DEVELOPMENT PLANNING

A broad Federal Development Plan (*Bundesraumordnungs-program*) is developed in greater detail through the Laender development plans, regional plans, and local development and land-use plans. An integrative approach is followed at all levels. While the federal development planning exercise focuses broadly on conservation of natural areas, planning at the local level (counties, districts) must integrate all relevant issues concerning biodiversity conservation.

The national strategy is expected to influence other national planning processes.

INTENDED TARGET OF PLANNING EFFORTS

The Parliament, the executive branch of the federal government, and the Laender governments were integrated into the participatory and review process of the strategy and are its prime target group.

RELATIONSHIP TO THE CONVENTION ON BIODIVERSITY

In light of the internationally agreed-upon language and structure of the Convention, the outline and structure of the German strategy paper closely followed each article of the Convention. This allows for comparison between and compatibility of different country's strategies and will facilitate and enhance the reporting process of the Convention.

SCALES OF PLANNING EFFORTS

The national strategy considers biodiversity at the ecosystem, species, and genetic levels, as well as factors that have an impact on biodiversity. The strategy addresses regional issues for areas of extraordinary importance—for example, the Wadden Sea. It also focuses on issues with bilateral importance (shared ecosystems), such as the Trilateral Wadden Sea Agreement (Germany, The Netherlands, Denmark); supranational importance, such as European regulations or agreements; and multi-lateral importance such as species under the Convention on International Trade in Endangered Species of Wild Fauna & Flora (CITES), the Convention on Migratory Species (CMS) or the Bern Convention, or areas under the Ramsar Convention, the World Heritage Convention, or the UNESCO Biosphere Reserve Program.

INTERNATIONAL ASSISTANCE

There was no financial support from international sources. Technical support was drawn from several sources, including UNEP—*Guidelines for National Biodiversity Strategies* and *Action Plans,* the *Global Biodiversity Strategy* of WRI, IUCN, and UNEP, and biodiversity strategies and action plans that have already been undertaken (e.g., in Australia, and the United Kingdom).

MONITORING AND FOLLOW-UP

The strategy will be monitored according to Article 26 of the Convention. Germany sees the strategy as a tool serving two purposes: to implement its biodiversity agenda and to report on a regular basis on its effectiveness in meeting the objectives of the Convention. The Conference of the Parties (COP) has yet to determine the intervals at which the contracting parties to the COP will report on the implementation process. Germany will follow these intervals and prepare an updated report by renewing the strategy accordingly.

The implementation of specific provisions by different levels of government will be monitored by the Federal Ministry for the Environment in order to gather information and identify gaps in the implementation process.

OBSTACLES AND CHALLENGES

There were several obstacles or challenges in the strategy process, including the lack of adequate personnel, potential problems concerning the willingness of the local government (Laender) to carry out nature conservation efforts, and insufficient financial resources. Scientific and technological problems may also arise out of the new and ambiguous formulations of technology transfer, agreements on access to genetic resources, and benefit-sharing agreements.

FACILITATING FACTORS

The ratification of the Convention by Parliament and the Laender were important driving powers for a national strategy. Unanimous support for ratification by the involved federal entities also facilitated the development of the strategy. The county-level governments, responsible for implementation, also support the strategy.

The formation of an expert Biodiversity Steering

Committee, the preparation of a Biodiversity Country Study and other documents on biodiversity management, and the formation of a national NGO forum on biodiversity were also instrumental.

COST, PERSONNEL, AND TIME REQUIREMENTS

Because the strategy was carried out by existing personnel on the federal level, no additional personnel costs are involved. The Ministry covers costs for personnel, printing, and translation.

Four people are working almost full-time on the strategy (one in the Ministry and three in the Federal Agency for Nature Conservation). Eight government departments with impacts on biodiversity have assigned at least one person at the federal level. At least one person per state is coordinating the Laender involvement. On the NGO side, the Biodiversity Working Group, with about 10 to 15 people, works on several aspects of the strategy.

The process will take approximately one year from preliminary discussions about a possible structure to the final version of the strategy in the English language.

4. Based on case study material prepared by Marc Auer, Executive Engineer, Federal Ministry for the Environment, Nature Conservation and Nuclear Safety, Bonn, Germany.

5. The Norwegian National Biodiversity Strategy[5]

INSTITUTIONAL BASE

The Directorate for Nature Management (DN), under the Ministry of the Environment, is the national executive authority on scientific and management-oriented questions related to biological diversity. In 1992, the DN prepared a country study on biological diversity (DN report 1992 5b). This was done in response to White Paper No. 46 on Environment and Development (1988–89), which called for information to aid in the preparation of a national biodiversity strategy.

In White Paper No. 13 (1992–93) to the Parliament (Storting), the Norwegian government stated that a national strategy and action plan would be prepared to implement the Convention on Biological Diversity. In accordance with the provisions of Article 6(b) of the Convention, the plan will be based on sectoral strategies drawn up by each of the ministries involved.

An interministerial committee representing nine ministries, led by the Ministry of the Environment, is responsible for preparing the strategy.

In July 1993, the Ministry of the Environment drew up guidelines for the content, organization, and results of the sectoral plans in close cooperation with the other ministries involved.

Sectoral plans have since been drawn up by the following ministries: Environment; Agriculture; Fisheries; Defense; Transport and Communication; Industry and Energy; and Education, Research and Church Affairs. Agencies within the various departments have been involved in preparing the plans. Within the Ministry of the Environment, the Directorate for Nature Management has been responsible for producing the draft. These plans were finalized and available for public comments in June 1994. The results of a planned public hearing were summarized in September 1994.

The various sectoral plans that may be revised as a result of the hearings will be included in a national strategy and action plan to be presented as a white paper to the government in March 1995.

METHODOLOGY

The country study was modeled on UNEP's *Guidelines for the Preparation of Country Studies on Costs, Benefits and Unmet Needs of Biological Diversity Conservation Within the Framework of the Planned Convention on Biological Diversity.* The study primarily describes the present state of knowledge about biological diversity (both wild and domesticated). It also reports on developmental trends, measures for conservation and sustainable use, economic value, and the need for research and analysis about biological diversity.

The sectoral plans are to be based on the provisions laid out in the Convention. The main elements of the plans are anticipated to be the identification of biological diversity and its components requiring conservation; identification of activities that threaten such diversity; protection and sustainable use of biological diversity, including impact analyses for major projects and important policies; monitoring of activities and processes that may have an adverse impact on diversity and conservation measures; and research, training, and information activities.

At the request of the Ministry of the Environment, eight of Norway's 435 municipalities are preparing plans for contributing to conservation and sustainable use of biological diversity. This work is also modeled along the guidelines set out for the various sectors and will be included in the national strategy for biological diversity.

PARTICIPATION

Based on White Paper No. 13 to the Parliament, the guidelines also emphasized the importance of ensuring that the plans enable all relevant sectors of society to gain access to information and opportunities to contribute to this work. The principle of open communication and community participation in decision making is emphasized in the chapters of *Agenda 21* dealing with the role of major groups and in the recently adopted Article 110(b) of the Norwegian Constitution, which deals with the right of the population to have access to information on environmental protection. To ensure that other authorities, business and industry, NGOs, and the general public have access to information and the opportunity to participate, public consultations on the sectoral plans will be held with all interested parties. Each ministry will use the results of these consultations to modify its draft sectoral plan.

GOALS AND OBJECTIVES

Norway's long-term objective concerning biological diversity is to preserve the productivity of nature and the diversity of species. This implies that important ecological processes and the natural basis for production must be maintained, while anthropogenic changes and potentially adverse impacts on the environment must be avoided—in particular, those that affect the genetic structure of natural populations.

To meet its objectives, Norway must increase its understanding of ways to maintain biological diversity and must play an active role in the international community to preserve biological diversity, develop sustainable use of resources, and promote fair and equitable sharing of benefits arising from those resources.

Each ministry is required to draw up an overall objective for its plan and define and discuss the problems involved in achieving this objective.

A first draft of the national strategy includes four objectives for the next five-year period:

- Education and public awareness shall be improved and form part of the basis for decision making about biodiversity.
- Land resources shall be used and managed to secure biological diversity in the short- and long-run.
- Biological resources shall be used and managed to maintain biological diversity in the short- and long-run.
- Activities which pollute and threaten biodiversity shall be minimized.

RELATIONSHIP TO NATIONAL DEVELOPMENT PLANNING

The biodiversity-planning exercise is especially related to two national processes. First, a systematic habitat conservation program based on regional inventories was initiated in Norway in the early 1970s. Regional conservation plans for each of the 18 counties in Norway gave priority to wetlands (especially those important for waterfowl), mires and bogs (primarily selected on botanical and hydrological criteria), rich deciduous forests (selected mainly on botanical criteria) and important seabird colonies. In 1985, inventories of coniferous forests began. Conservation planning for coniferous forests has been given high priority since 1988, on the recommendations of a national task force on their protection. A new national park plan adopted in 1993 will offer protection to 20 new national parks, enlargement of nine existing national parks, 16 new landscape-protected areas, and one large nature reserve covering a total of about 20,000 sq. km. These protected areas will be established within the next 10 years.

Second, White Paper No. 46 (1988–89) on Environment and Development states the government's strong emphasis on incorporating the principles of sustainable development in all national planning and sectoral policies. Agricultural, fishery, energy, transport, and other authorities must ensure that development and planning within their sectors are consistent with these principles of development and that budgetary as well as other means (e.g., administrative, judicial) are developed to reduce and avoid new environmental problems.

INTENDED TARGET OF PLANNING EFFORTS

The biodiversity-planning process directly involved those responsible for making decisions and advising the various ministers on policy. The national strategy will provide overall policy guidance for all ministries and their sector agencies at all levels.

RELATIONSHIP TO THE CONVENTION ON BIODIVERSITY

The biodiversity-planning process addressed the articles of the Convention explicitly. Both the guidelines for the sectoral plans and the sectoral plans themselves refer to the articles of the Convention. The sectoral plans also state objectives and actions in keeping with the Convention. However, neither the sectoral plans nor the draft strategy are structured article by article.

SCALES OF PLANNING EFFORTS

Both the sectoral and municipal approaches to producing the national biodiversity strategy address national, regional, and local issues. Following up on the objectives will require local and regional actions in particular.

INTERNATIONAL ASSISTANCE

Norway has not received technical or financial support from international sources for the development of the biodiversity strategy.

MONITORING AND FOLLOW-UP

Monitoring, reporting, and evaluating processes are important parts of the strategy. Reporting and evaluation at five-year intervals will be finalized in the national strategy.

OBSTACLES AND CHALLENGES

There have been remarkably few obstacles in producing the biodiversity study and the sectoral plans. However, despite the substantial amount of knowledge already available, the extent of the strategy makes lack of knowledge a constantly recurring problem.

A challenge for the future is to truly integrate environmental and biodiversity concerns into all sectoral policy and budgetary work as stated in White Paper No. 46.

FACILITATING FACTORS

Existing legislation, policies, and programs have facilitated the development of the strategy, as have the policy-development expertise of those involved, the preparation afforded by UNCED, and existing biodiversity conservation activity.

COST

Cost estimates are not available at this time.

PLAN

The various sectoral plans became available to the public on June 1, 1994. After broad public consultation with organizations, universities, research institutions, ministries and other interested parties, a summary of public opinions was prepared in September. The various sectoral plans have after that been revised and are now being integrated into a national strategy and action plan. This will be presented to the Parliament in autumn 1995.

5. Based on case study material prepared by Peter Schei, Director, Directorate for Nature Management, Trondheim, Norway; and Gudrun Schneider, Ministry of Environment, Oslo, Norway.

■ 6. *The Indonesian National Biodiversity Strategy, Action Plan, and Country Study*[6]

By 1994, Indonesia had completed a *Country Study on Biological Diversity* (GOI, 1992) under the auspices of UNEP, a *National Strategy for Biodiversity Management* spearheaded by KLH (Ministry of State for Population and Environment—GOI, 1993c), and the *Biodiversity Action Plan for Indonesia,* led by the National Development Planning Board (Bappenas) with financial assistance from the World Bank (GOI, 1993a).

INSTITUTIONAL BASE

The Directorate General of Forest Protection and Nature Conservation (PHPA) within the Ministry of Forestry is responsible for the protection of natural habitats. The Ministry of State for Population and Environment (KLH)—which became the Ministry of State for Environment (LH) in 1993—has played a larger role in national biodiversity planning. From its formation in 1978, KLH, in cooperation with the Indonesian Sciences Institute (LIPI), has been at the forefront of explicit thinking about biodiversity conservation. KLH was also instrumental in the development of the National Conservation Strategy in the early 1980s.

KLH was the lead agency in carrying out the *Indonesian Country Study on Biological Diversity,* prepared in early 1992 under the auspices of UNEP. The study was written largely by a consultant, under the guidance of a Biodiversity Country Study Standing Committee. Drawing on the draft *Biodiversity Action Plan* for most of its data on biodiversity status and trends, the study then quantitatively analyzes the benefits and costs of biodiversity management and the unmet financial needs.

KLH has played a very important role in articulating the breadth of the biodiversity issue for Indonesia and in building understanding of its international dimensions. However, it is a coordinating ministry, with no "line" authority or institutional presence on the ground. It does not have a decisive role in national development planning, either, although its recommendations in the environmental field are influential.

Thus, when the World Bank began discussions in 1990 about assisting Indonesia in developing a national biodiversity action plan, work gravitated to Bappenas, the ministry charged with all aspects of national planning and, importantly, budgeting. Bappenas had been involved in natural resource protection for some time through its planning mandate, and it was an important actor in establishing KLH in the 1970s. Since 1990, Bappenas has taken the lead in national biodiversity planning.

METHODOLOGY AND PARTICIPATION IN THE NATIONAL BIODIVERSITY ACTION PLAN

Participation in the development of the action plan departed from that of similar planning exercises in Indonesia. The action plan's development was guided by an intersectoral steering committee chaired by Bappenas and with members from KLH, Agriculture, Forestry, Domestic Affairs, LIPI, universities, and professional organizations such as the Indonesian Wildlife Society, Indonesian Forum on the Environment (WALHI), and Indonesian Rainforest Action Group (SKEPHI). Much of the action plan was drafted by foreign consultants who have long experience in Indonesian biodiversity issues.

Once a draft was prepared in both Bahasa Indonesian and English, Bappenas organized a three-day workshop to review it, in collaboration with the Bogor Agricultural Institute and with the technical and financial support of the World Bank and the government of Norway, as well as participation from several international NGOs, including IUCN, WRI, and WWF. The meeting brought together over 50 people from relevant government agencies, NGOs, the scientific community, and international donors. Notably, both provincial government officials and local NGOs based outside Jakarta were present.

The steering committee and the consultants used the results of the meeting plus written comments to revise the draft over the next several months. By July 1991, a final draft was widely circulated. It took two more years for the final version to be produced in English, and the Bahasa Indonesian version is due in mid-1994 or later with full government approval.

GOALS AND OBJECTIVES

The primary objective of the action plan is to catalyze immediate action to slow the rate of biodiversity loss and to develop a strategy that allows sustainable utilization of natural resources while conserving biodiversity and the natural resource base. The action plan therefore provides an integrated operational framework to set priorities and guide investments.

Specifically, the action plan lists three general objectives:

- to slow the loss of primary forests, wetlands, coral reefs, and other terrestrial and marine habitats of primary importance for biodiversity
- to expand the data and information available on the nation's biodiversity and make it available to policy makers and the public, and
- to foster the utilization of biological resources in ways that are sustainable and less harmful than current practices.

RELATIONSHIP TO NATIONAL DEVELOPMENT PLANNING

The action plan has substantially influenced Indonesia's Sixth Five-Year Development Plan

(1994/95–1998/99), which includes sections on affirmative conservation action. Biodiversity has been increasingly addressed through the mainstream development process directed by the National Development Planning Agency (Bappenas). Bappenas has issued written instructions to all of its planning staff to follow the action plan. In addition, some believe the action plan can be used to monitor and, in some cases, to change sectoral development proposals.

INTENDED TARGET OF PLANNING EFFORTS

The document is clearly targeted at policy makers within the Indonesian government and donors. As an official government document, it provides an authoritative framework to use in planning, strengthening institutions, policy reform and development, priority setting, and project choice and design. Its Bahasa Indonesia version should be of use as a reference tool for the academic, NGO, and scientific community, provided it is widely distributed.

RELATIONSHIP TO THE CONVENTION ON BIOLOGICAL DIVERSITY

Because both the strategy and the action plan were developed before the Convention on Biological Diversity was finalized, neither refers specifically to the Convention. Indonesia's active participation in the Convention negotiations, its negotiators' close involvement with the biodiversity-planning processes at home, and the participation of many key Indonesians in the *Global Biodiversity Strategy* process, has ensured that the action plan responds to many of the Convention's provisions.

Indonesia is currently in the process of ratifying the Convention through promulgation of a Basic Law on Biodiversity. It is likely that the Law will also mention the national strategy and/or action plan, giving them further authority.

SCALES OF PLANNING EFFORTS

Indonesia is involved in discussions with both Malaysia and Papua New Guinea concerning the establishment of trans-frontier protected areas, it was active in the Biodiversity Convention negotiations, and it is a member (or prospective member, in the case of the Ramsar Convention) of several regional and international conservation agreements. However, the action plan did not delve into regional or international agreements and cooperation. Rather, the process focused on bringing more of the provincial and local perspective into the dialogue.

INTERNATIONAL ASSISTANCE

The action plan and the national strategy were domestically driven processes, and international donor assistance built on and augmented processes that were already under way in one form or another. The World Bank assisted in the preparation of the action plan with a technical-assistance grant. Bank personnel were also involved in the process as advisors and catalysts. IUCN, WRI, and the WWF-Indonesia Program were all involved as advisors and technical resources. The Government of Norway supported preparation of the country study, and the government of Canada has provided general support for KLH's biodiversity work.

MONITORING AND FOLLOW-UP

Both the strategy and the action plan call for the establishment of a National Biodiversity Commission to coordinate and monitor follow-up. The action plan contains a separate chapter on the "Strategy for Implementation of the Plan," which provides details on steps to be taken, including the establishment of a review process to provide indicators and assessments of progress. It also notes that "the Biodiversity Commission should carry out regular evaluation and reviews, recommend follow-up actions and decide on future 'programs' for action, based on priorities outlined in the Plan" (p. 59).

Bappenas has established a five-person review commission to monitor the implementation of the action plan and to screen projects in all sectors with regard to their impacts on biodiversity and their adherence to the priorities and guidelines laid out in the action plan.

OBSTACLES

Obstacles to the planning process included the inherent difficulty of coordinating the numerous sectors and interests concerned with biodiversity, turf battles over which agency should lead biodiversity-planning efforts, and, more seriously, the difficulty of bringing provincial and local perspectives into the planning. The private sector has also been largely absent from the process. The institutional weakness of the Environment Ministry and the PHPA are key obstacles to implementation of the action plan.

FACILITATING FACTORS

The Ministry of Environment, LIPI, and Bappenas all exerted strong leadership in the process. The process was also expedited by pressures from the activities preceding UNCED, from international and local NGOs' participation in the *Global Biodiversity Strategy* process, and by strategic donor assistance. The availability of international expertise was also helpful. Finally, by welcoming national NGOs into the process, Indonesia gained substantive input and defused a potentially sensitive political flank both nationally and internationally.

RESULTS TO DATE

Because Indonesia's national biodiversity-planning efforts are well-developed and in an advanced stage, there has been some progress in implementing the action plan. The action plan and associated developments have been beneficial for PHPA's budget, which was increased fivefold in 1994 from the previous year. Various sectoral agencies are now formulating their own responses to the plan for submission to Bappenas and, if accepted, into annual budgets and work plans.

The action plan is a useful tool in working with aid donors because it gives them a set of national priorities around which to develop their biodiversity programs. The plan also gives the government the ability to screen donor ideas for those that really match the government's national priorities for biodiversity conservation.

Biodiversity conservation activities have greatly increased in the three years since the draft action plan was produced. Not all of these are directly attributable to the strategy or action plan, but they have certainly played an important catalytic role.

6. Based on case study material prepared by Charles Barber, Senior Associate, World Resources Institute, through interviews with Suraya Afiff (WALHI), Hadi Alikodra (Indonesian Ministry of State for Environment), Jerry Bisson (USAID-Indonesia), Herman Haeruman (Indonesia National Development Planning Agency), Kathy MacKinnon (Environment Department, World Bank; formerly of WWF-Indonesia), Agus Purnomo (Pelangi Indonesia Institute), Mien Rifai (Herbarium Bogoriense, Indonesian Institute of Sciences), Effendy Sumardja (Indonesian Ministry of Forestry), R.E. Soeriaatmadja (Indonesian Ministry of State for Environment), R. Yusuf (Indonesian Ministry of Forestry), and Arief Yuwono (Indonesian Ministry of State for Environment).

■ 7. *The United Kingdom Biodiversity Action Plan*[7]

INSTITUTIONAL BASE

Biodiversity: The UK Action Plan was published and launched by the prime minister in January 1994. The document represents the first United Kingdom Biodiversity Action Plan. Because of the UK's tradition and science base over 200 years, it was decided to include the country study, the strategy, and the action plan in one document.

The action plan was managed by a steering group chaired by the Department of the Environment. The group included representatives of the territorial organizations (Scottish Office, Department of Environment (Northern Ireland), and Welsh Office); government departments such as the Treasury, Foreign and Commonwealth Office, Ministry of Agriculture, Fisheries and Food, and the Forestry Commission; and the nature conservation agencies English Nature, Scottish Natural Heritage, Countryside Council for Wales, Joint Nature Conservation Committee, Countryside Commission, and the Natural Environmental Research Council.

The plan was approved by ministers of all the relevant departments before it was published and was fully endorsed by the Central Government.

METHODOLOGY

The action plan follows guidelines established by the UN Environment Programme for Country Studies and the provisions of the Convention.

Contributions were made by delegates representing all sectors of society. The main stages were as follows:

- the creation of an interdepartmental or interagency steering group
- the preparation of an outline for the plan
- the issuing of a requesting-consultation letter from the national government agencies and biological collections, university departments, industry and commerce, and voluntary conservation organizations and individuals
- a two-day seminar at the Royal Geographical Society in London attended by over 100 delegates representing different sectors
- the appointment of chapter editors drawn from government, agencies, collections, academia

and voluntary conservation organizations
- central editing, then design and publication, and
- launch by the prime minister and ministers from the relevant government departments.

PARTICIPATION

The action plan's editorial team was supported by a small subcommittee that provided scientific and technical assistance.

GOALS AND OBJECTIVES

The goal of the plan was to conserve and enhance biological diversity within the United Kingdom and to contribute to the conservation of global biodiversity through all appropriate mechanisms. The objectives were:

- to conserve and, where appropriate, to enhance the populations and natural ranges of all species native to the UK and the quality and range of their habitats and ecosystems; internationally important and threatened species, habitats, and ecosystems; locally important and threatened species, habitats, and natural and managed ecosystems; and the biodiversity of all natural and seminatural habitats;
- to increase public awareness of, and involvement in, conserving biodiversity; and
- to contribute to the conservation of biodiversity on a European and global scale.

INTENDED TARGET FOR PLANNING EFFORTS

As well as reaffirming policies and programs, the action plan contains 59 specific tasks and actions. These provide overall policy guidance for government departments and agencies with jurisdiction over conservation activities.

RELATIONSHIP TO THE CONVENTION ON BIODIVERSITY

The chapters of the plan are based on articles of the Convention.

SCALES OF PLANNING EFFORTS

The action plan concentrates on national wildlife. Although it also addresses regional and local issues, it emphasizes the role of community groups in local conservation efforts. The Biodiversity Action Plan steering group will prepare and implement a campaign to increase public awareness and involvement in conserving UK biodiversity.

The action plan also addresses UK support of biodiversity overseas, which is instituted in part by CITES, the Darwin Initiative, and overseas development programs.

INTERNATIONAL ASSISTANCE

The costs of preparing and publishing the plan were paid for by the UK government.

MONITORING AND FOLLOW-UP

A new Biodiversity Action Plan steering group has been set up for a two-year period to accomplish the following tasks:

- the development of specific goals for key species and habitats for the years 2000 and 2010, including the cost of achieving those goals
- the creation of a working group to improve the accessibility and coordination of existing biological data sets, and
- the establishment of a review process for the success of the action plan.

OBSTACLES

The scope of the task and the short time frame in which it had to be accomplished were the only obstacles.

FACILITATING FACTORS

The enthusiasm and commitment of the many contributors and the guidance of the interdepartmental, interagency steering group made the development of the Action Plan possible. There was substantial agreement over major issues.

The action plan has generated interest: its progress post-Rio has been debated in the House of Commons and the House of Lords; it has been reviewed extensively in the press and media; and ministers have discussed its contents on television and radio. Biodiversity has become an increasingly popular theme at conferences and seminars.

COST, PERSONNEL, AND TIME REQUIREMENTS

The UK government financed the process, and the tight timetable was agreed upon by the prime minister to demonstrate the country's commitment to the Convention. Although figures cannot be provided on costs, approximately 52 people (two full-time) worked on the document, and dozens of others provided comments throughout the process.

7. *Based on case study material prepared by Roger Bendall, European Wildlife Division, Department of the Environment, Bristol, United Kingdom.*

■ 8. *The Vietnamese Biodiversity Action Plan*[8]

INSTITUTIONAL BASE

The Biodiversity Action Plan for Vietnam was prepared under the leadership of the Ministry of Science, Technology and Environment, in cooperation with universities, research institutes, and other ministries (primarily the Ministry of Forestry). It was prepared as part of a GEF (UNDP) biodiversity project for which the lead consulting agency was World Wildlife Fund. The World Conservation Union (IUCN) was a subcontractor in the project. The project was approved by the Council of Ministers and the State Planning Committee, the highest decision-making bodies in Vietnam.

METHODOLOGY

Leading Vietnamese scientists and members of the IUCN Secretariat met in Hanoi on February 22-23, 1993, to initiate an action plan. Sub groups discussed biodiversity in terrestrial and freshwater ecosystems; biodiversity in marine and coastal ecosystems; the policy and legal framework for biodiversity; and human impacts on biodiversity.

After the initial meeting, the action plan team carried out field investigations in 12 provinces of the country, assessing various issues by region. The action plan itself was written by a foreign consultant with many years of experience in Vietnam who worked in close collaboration with Vietnamese colleagues on the team. IUCN provided a team leader and a marine biologist.

Vietnam has not prepared a UNEP biodiversity country study, but through IUCN and WWF, a national conservation strategy was prepared in 1985. The strategy provided a firm foundation for the action plan.

PARTICIPATION

Several institutions selected for their expertise contributed or were consulted. These include the Ministries of Fisheries, Energy, Sea Products, Education, Public Health, Forestry, Agriculture, and Science, Technology, and Environment and the Institutes of Oceanography, Ecology and Biological Research, Aquatic Products and Economic Ecology. Also participating were the Centre of Ecology, the State Committee for Environment and Sustainable Development, the Forest Inventory and Planning Institute, the Department of Protection of Aquatic Resources, the Mangrove Ecosystem Research Centre, the Haiphong Marine Research Centre, the Nha Trang Marine Institute, the University of Hanoi, the Centre for Resource Management and Environmental Studies, the Council of Ministers, and the State Planning Committee.Those government agencies expected to implement the plan were directly involved in its preparation. Because Vietnam has few NGOs *per se*, these were not involved.

GOALS AND OBJECTIVES

The objectives of the Biodiversity Action Plan are:

- to guide all domestic endeavors in the field of biodiversity conservation down to provincial-level actions
- to enable international agencies to understand what the priorities for investment are in Vietnam, and
- to convert the policy contained in the National Plan for Environment and Sustainable Development into a detailed program of action.

In addition, the goals of Vietnam's Strategy for the Preservation of Valuable Biological Resources are:

- to protect examples of all major ecosystems *in situ* through a nationwide system of protected areas
- develop sustainable utilization of living resources, including forests, and
- to provide specific *ex situ* conservation measures for species when necessary.

RELATIONSHIP TO NATIONAL DEVELOPMENT PLANNING

The action plan was designed to help implement the National Plan for Environment and Sustainable Development 1991-2000.

INTENDED TARGET OF PLANNING EFFORTS

The action plan has already provided considerable policy guidance for decision making at national and provincial levels, including investment priorities. The action plan calls for the establishment of a National Biodiversity Authority, to be part of the State Committee for the Environment and Sustainable Development.

RELATIONSHIP TO THE CONVENTION ON BIOLOGICAL DIVERSITY

The action plan makes specific recommendations for the implementation of each of the key articles of the Convention.

SCALES OF PLANNING EFFORTS

The preparation of the action plan addressed the provincial level, involving surveys and questionnaires in 12 provinces, and included the design of specific projects for implementation. It specifically identified transfrontier protected areas (bordering with Cambodia, Laos, and China) where international cooperation would be useful. The action plan also specifically addressed the requirements for implementing CITES, which Vietnam is in the process of joining.

INTERNATIONAL ASSISTANCE

The preparation of the action plan received financial support from GEF, was implemented by WWF with technical support from IUCN, and was coordinated by an international consultant.

MONITORING AND FOLLOW-UP

UNDP is the funding and coordinating agency for the Vietnam GEF project. The action plan proposes a biodiversity monitoring program to determine whether application of improved management and regulations is having a positive effect on biodiversity conservation. This monitoring program would have six elements: habitat, protected areas, indicator species, data, management, fisheries stocks, and physical parameters.

OBSTACLES

The shortage of trained Vietnamese scientists and managers who could participate actively in the planning team, especially at the provincial level, was a major obstacle. The training component of the project is designed to increase the number and effectiveness of staff.

Major constraints against conserving biodiversity that were identified in the action plan include lack of funds, low awareness by government agencies and the public, gaps in knowledge, shortages of equipment and trained staff, and the absence of clear regulations, instructions, and adequate laws. All but one province considered lack of funds to be the biggest obstacle.

The action plan has had insufficient input from economists, so the relationship between fiscal policy and biodiversity is hardly covered. This may prove to be a significant limitation.

FACILITATING FACTORS

The preparation of the action plan was greatly facilitated by the preparation of earlier plans, as discussed above, and a series of field research expeditions throughout the country over the past several years.

The country has a good set of vegetation maps, a good review of major wetlands, detailed priorities for action in major forest reserves, an understanding of the distribution of key species of wildlife, and a detailed protected-areas-system plan. Field studies have also identified rankings of conservation values and threats to coral reefs.

The action plan is now in the form of an advanced draft but has not yet been formally submitted to government. However, even in draft form, the action plan is being used to help assess priorities for investments. It includes some 52 project concepts to address key actions, calling for a total investment of US$876 million.

COST, PERSONNEL, AND TIME REQUIREMENTS

The process of preparing the action plan is still continuing, but its preparation to date has involved several dozen people directly and several hundred more indirectly.

The cost of preparing the Biodiversity Action Plan is US$218,000 to date, paid for as part of a US$3 million project on "Conservation Training and Biodiversity Action Plan" funded by GEF through the United Nations Development Programme.

8. Based on case study material prepared by Jeffrey McNeely, Chief Biodiversity Officer, World Conservation Union (IUCN), Gland, Switzerland.

■ 9. The Dutch National Biodiversity Action Plan[9]

INSTITUTIONAL BASE

In 1990, the Dutch government published three major policy plans on the natural environment in line with the Brundtland Commission on Sustainable Development: the Nature Policy Plan, the Environmental Policy Plan, and the Third Document on Water Management. These form the main body of the national strategy on biodiversity conservation. Additional sector policies, such as those on forestry, agriculture, and fisheries, provide a further basis for sustainable use of biological resources.

The Nature Policy Plan, published by the Minister for Agriculture, Nature Management and Fisheries, is the most important document in the strategy, with policy on the sustainable conservation, rehabilitation, and development of natural resources. Within the Ministry, the Department for Nature, Forests, Landscape and Wildlife coordinates nature conservation policy. A strategic action plan is being prepared to complement these documents.

METHODOLOGY

As indicated earlier in this report, the Netherlands prepared a strategy for biological diversity before the Convention was signed. The Convention, and to a certain extent *Agenda 21*, led to a process of further elaboration of this strategy, by use of the following systematic methodology.

Step 1: Formal analysis

Thirteen relevant Dutch policies were analyzed for gaps in, and the extent of, their formal compliance with the provisions of the Convention on Biological Diversity: policies on nature, environment, physical planning, water management, agriculture, forestry, fishery, education, research, technology, recreation, international trade and environment and development cooperation.

This step has been implemented by the consulting firm AIDEnvironment, under the auspices of the Ministry of Agriculture, Nature Management and Fisheries and in close consultation with all Departments concerned.

Step 2: Broad, informal analysis

The Convention on Biological Diversity, in conjunction with *Agenda 21*, was analyzed for potential new policies for The Netherlands related to biological diversity. The focus was on generating ideas and possibilities in all relevant policy sectors, not on political feasibility. For the sake of the analysis, the Convention and *Agenda 21* were summarized into 100 key questions, which were then applied to the policies mentioned to identify gaps in coverage.

Step 3: Identification of main themes and spearhead actions

Main themes for new policy have been identified and elaborated into novel but pragmatic spearhead actions. These were discussed during a one-day workshop in June 1994. Participants at the workshop included all relevant government departments and semi-private and private research, information and expert institutes. Key economic sectors, environment and nature conservation NGOs, and individual experts also participated.

The preliminary choice of main themes and spearhead actions includes:

- Strengthening the conservation and sustainable use of biological diversity in nature and environment policies. Actions should include strengthened policies for biological diversity outside protected areas.
- Integration of conservation of biological diversity into other relevant policy sectors, such as agriculture, fisheries, and forests.
- Strengthening of research, assessment, and monitoring with regard to policy development for conservation and for prevention of adverse external processes and influences. Actions should include inventory, coordination and further planning of research, and ongoing assessment and monitoring of conservation efforts.
- Participation in international policy on biological

conservation. Actions should include development cooperation (a specific sector policy document on biological diversity and development cooperation is in an advanced stage of preparation) and external integration (integration of nature conservation and environment policies into other international policy sectors, e.g., with those of the European Union).

- Development of a mechanism for a technological information clearing house.
- Provision of public information. Action should include the development and use of instruments for educating the public about the biological diversity agenda.

Step 4: Compilation of the results into a complementary strategic plan of action

The results of the workshop will contribute to finalizing the choices for main themes and spearhead actions. They will be elaborated into a coherent complementary strategic plan of action. This will be presented to the Parliament and implemented by the respective ministries.

PARTICIPATION

The preliminary concept of the Nature Policy Plan was discussed with 33 major NGOs and advisory councils in the field of nature conservation and environmental protection. The general public was also invited to comment. Official meetings were organized with provincial authorities and representatives of municipalities and regional water boards.

GOALS AND OBJECTIVES

The overall goal of the Nature Policy Plan is the sustainable conservation, rehabilitation, and development of natural resources, both in The Netherlands and abroad. The operational objective was to intensify and expand existing efforts in the field of nature conservation, building on the feeling that concrete actions are urgently needed. This resulted in some new policy instruments, particularly in nature development, and a doubling of the budget (up to about DFL 350 million annually). At the same time, environmental policy that focuses on sustainability issues was extended and intensified by the Environmental Policy Plan.

RELATIONSHIP TO NATIONAL DEVELOPMENT PLANNING

The Nature Policy Plan is a national government plan. By now, the Nature Conservation Act has been adapted to create a legal basis for the plan.

Many other policy documents relate to biological diversity planning, including:

- physical planning of the country's "Green Space"
- agriculture (including sustainability of agriculture)
- forestry, including both a National Forest Policy Plan and a Governmental Position on Tropical Rainforests
- fishery, including a policy for sustainable coastal and sea fisheries
- development cooperation: the general policy frameworks contain specific provisions on the conservation of biological diversity, and explicit policy-sector document on biological diversity and development cooperation is in an advanced stage of preparation.

Other policy areas that contain explicit or implicit provisions on biological diversity are: education, research, technology, recreation, and international trade and environment.

INTENDED TARGET OF PLANNING EFFORTS

The Nature Policy Plan explicitly provides guidance for decision makers within agriculture, forestry, fisheries, hunting, infrastructure works, and excavations, among other sectors. The plan stresses that provincial authorities should play a key role in the realization of the plan, and for this, a contract was signed between the Minister of Agriculture and the 12 official provincial representatives.

RELATIONSHIP TO THE CONVENTION ON BIODIVERSITY

When most policies were being defined, two years before UNCED, it was not yet possible to anticipate the exact results laid down in the Convention. However, compliance of the respective Dutch policies has turned out to be satisfactory despite several gaps and the need for further or intensified actions.

SCALES OF PLANNING EFFORTS

The Nature Policy Plan's major focus is on habitat protection, the basis for the national ecological network and species policy. The plan states that biological diversity should be the aim under the most natural conditions. The national planning concept has been used by provincial and local authorities, almost all of whom have adapted their (physical) plans to take the natural environment into account.

The plan also concerns international nature-conservation policy. It uses the same idea of national habitat protection within an ecological network. It also encompasses additional species policies aligned with the nature-conservation activities of the European Union (Habitats Directive, which also includes provisions for the protection of wild species of flora and fauna) and the Council of Europe (European Ecological Network).

Special attention is given to neighboring countries' trans-boundary natural areas, the Wadden sea and the large Rhine, Meuse, and Scheldt rivers. For these ecosystems, separate international boards are active. Efforts are also under way to promote a Biological Diversity Strategy on a European scale.

INTERNATIONAL ASSISTANCE

International assistance was not used to formulate the Nature Policy Plan. But the Dutch government is supporting the IUCN nature-policy planning process in several Eastern European countries: Hungary, Poland, the Czech Republic, and the Slovak Republic. On a bilateral basis, the Dutch government is supporting processes in Russia and the Ukraine. By now, several countries have prepared or are preparing similar plans, and the concept of ecological networks has become widespread.

The Dutch government also supports the Netherlands Antilles' efforts to write a nature policy plan. The director general of Development Cooperation has a specific budget line for biological diversity in developing countries. Part of the funds is channeled through the GEF, and another major part is channeled through bilateral cooperative programs.

MONITORING AND FOLLOW-UP

Each year, a multi-year program is sent by the Minister of Agriculture, Nature Management and Fisheries to the parliament, presenting an overview of the extent to which the aims of the Nature Policy Plan are being realized. Instruments, projects, and budgets are outlined in detail.

A network of data collection and monitoring of field results has been established, coordinated by the Information and Knowledge Centre on Nature, Forests, Landscape and Wildlife of the Ministry for Agriculture, Nature Management and Fisheries. Cooperating are the Ministry for the Environment, the Ministry for Water Management, provincial authorities, managers of nature reserves, Bird Protection, and volunteers. Every four to five years, the results are published in the *State of Nature*.

OBSTACLES

Development of the Nature Policy Plan did not meet many obstacles. Support for nature and the environment was, and still is, strong in The Netherlands. A major problem is that although budgets have been increased tremendously on an annual basis, much more money is needed in the first decade of realization to prevent further decline of habitats and species. At the same time, part of the additional budgets was transferred to the provincial authorities. The short-term funding problem has been overcome by creating a fund for nature conservation policy, putting isolated budgets together and thus creating flexibility in time and space.

The designation of nature preserves was, and will continue to be, difficult. Natural areas have declined in size and become highly fragmented. It is necessary to expand the remaining areas by turning 50,000 hectares of agricultural land into nature reserves. The turmoil in society resulting from this reverse trend can be overcome by demonstrating that this has no harmful effect on agricultural production. Now, individual farmers are being asked to sell part of their land to the government for conversion to nature reserves.

FACILITATING FACTORS

The worldwide support for nature conservation and environmental protection and the urgency to take action are the most important facilitators in The Netherlands. Both subjects are still high on the political agenda, though slightly less so than they were a few years ago. The Nature Policy Plan is fully supported by the Dutch Parliament. Also, at the provincial and local levels, actions based on the Nature Policy Plan are strongly facilitated by this support.

A series of background studies on the alarming state of nature and the environment in the Netherlands have also helped facilitate biodiversity-conservation actions. In particular, a document on the international importance of Dutch biological resources has illustrated that the survival of such common species as meadow birds is highly dependent on protection measures taken within The Netherlands.

COST, PERSONNEL, AND TIME REQUIREMENTS

The development of the Nature Policy Plan took about two years with the full-time participation of three to four senior officers, supported, when needed, by national officers working in nature conservation. Additional costs were incurred for publication and information activities: 15,000 copies of the plan were produced, plus 7,000 copies of an English language edition and brochures in English, German, French, and Spanish. The estimated total costs are about DFL 1 million (only a fraction of the annual budget for nature conservation of DFL 350 million). The Action Plan is the responsibility of a small team from the Ministries of Agriculture and Nature Management, and Environment.

9. Based on case study material prepared by Marcel L. Vernooy, Ministry of Agriculture, Nature Management and Fisheries, Department of Nature, Forests, Landscape and Wildlife, Division of International Affairs, The Netherlands, and Bart Romijn, AIDEnvironment, The Hague and Amsterdam, resp., The Netherlands.

■ 10. The Chinese Biodiversity Action Plan[10]

INSTITUTIONAL BASE

As a part of actions preceding the Biodiversity Convention, the Biodiversity Conservation Action Plan became Part A of the GEF China Biodiversity Project.

The State Environment Protection Committee (SEPC), under the auspices of the State Council, coordinates efforts to solve important environmental issues in China by providing guiding principles and evaluating relevant laws and policies. Because the National Environmental Protection Agency (NEPA) is responsible for managing and evaluating biodiversity conservation, it was appointed to be the lead agency in coordinating efforts for drafting the Biodiversity Conservation Action Plan (BAP). The Chinese Academy of Sciences (CAS) was also heavily involved in the planning process. The Ministry of Forestry, Ministry of Agriculture, State Oceanic Administration, and Ministry of Construction are responsible for administration, management and research work on biodiversity conservation, with differing emphases and concerns. In addition, the State Science and Technology Commission, State Planning Commission, and Ministry of Finance participated in the planning process and will be the main domestic financial resources for the implementation and enforcement of the Biodiversity Action Plan. An international advisory group was also formed to oversee the drafting of the plan.

METHODOLOGY

The preparation of the Action Plan was guided by the Convention on Biological Diversity. Numerous other international documents were also used as important references. The expert team based the plan's outline and the evaluation of present status and needs primarily on the Convention. The international advisory group and the World Bank's project manager were also consulted. The participating institutions then worked out tentative action plans, paying special attention to their function and responsibilities. After a series of symposia and workshops, the sectoral action plans were compiled into the first draft of the national Action Plan. This document was distributed to relevant institutions and interested parties for comments.

A biodiversity country study is tentatively being planned for China as a UNEP project. One main issue to be resolved is whether the Biodiversity Country Study will cover a wider scope than the Action Plan while giving special attention to integrating conservation with sustainable development, especially local development.

PARTICIPATION

NEPA had the lead role in organizing, coordinating, and drafting the Action Plan. Numerous other institutions took part, including the CAS, several ministries, state administrations, and commissions.

- The Chinese Academy of Sciences: CAS plays the leading role in the study of biological resources. It possesses the most comprehensive information on this topic and provides scientific consultation to the State Council.
- Ministry involvement: The Ministry of Forestry manages the natural reserves, governs the import and export of flora and fauna, implements the Wildlife Protection Law, and formulates regulations for conserving forest ecosystems. The Ministry of Agriculture is in charge of protecting agricultural and grassland ecosystems and freshwater and marine fisheries as well as other aquatic species. The Ministry of Finance allocates funds for biodiversity conservation activities. The Ministry of Public Security assists in implementing Wildlife Protection Law and Environment Law. It also acts in enforcing laws regarding illegal trade, hunting, and destruction of wildlife and habitats.
- Administrations and commissions: The State Oceanic Administration oversees the management and use of the marine ecosystem. The State Planning Commission incorporates biodiversity conservation into the state annual development plans and long-term strategies. The State Science and Technology Commission is in charge of science policy and scientific research activities at the state level.

Although the Ministry of Construction oversees zoos, botanic gardens, and national parks and plays an important role in both *in situ* and *ex situ* protection and captive breeding programs, only representatives of *ex situ* conservation participated in the planning. No representatives from the Ministry's department of national parks took part.

The ideas and opinions of university-based scientists and grassroots groups were also collected and are reflected in the Action Plan. The above organizations presented their ideas and comments at appropriate stages during the whole process of preparation and revision, and no group was excluded.

GOALS AND OBJECTIVES

The goals of the Biodiversity Conservation Action Plan are:

- set priorities and identify feasible measures to stop the destruction and loss of biodiversity and habitats;
- over the long term, save endangered species, conserve living resources, use natural resources rationally and sustainably, and restore ecosystems as much as possible; and
- offer scientific assistance in rural development such that it agrees with biodiversity conservation.

The overall goal and operational objectives were determined through intensive discussions among scientists and representatives from participating organizations.

RELATIONSHIP TO NATIONAL DEVELOPMENT PLANNING

As one of the follow-up actions of UNCED, the State Science and Technology Commission (SSTC) and the State Planning Commission (SPC) were designated to organize and draw up the "Agenda 21 for China." In selecting the country's major five-year research projects, the SSTC and SPC, major funding bodies for long term research and development programs, considered the actions of biodiversity conservation suggested in the Action Plan. The work of biodiversity information management, conservation and sustainable use of biodiversity, the preservation of genetic diversity, and *ex situ* conservation of endangered plants has already been conducted by the key research programs of the CAS's Eighth Five-year Plan. It is believed that the programs will be further strengthened under the Ninth Five-year Plan.

INTENDED TARGET OF PLANNING EFFORTS

The Action Plan proposed a national biodiversity monitoring network that was asked to prepare periodic status reports for government and the public.

RELATIONSHIP TO THE CONVENTION ON BIODIVERSITY

The completion of the Action Plan is a step in the implementation of the Convention on Biological Diversity. The Convention was used as a guide in drafting the Plan.

SCALES OF PLANNING EFFORTS

The Action Plan deals mainly with issues on the national scale. Regional issues were addressed in areas of high ecological importance, such as those areas where species are highly indigenous or rich, or where there are flagship species. There was no cooperation with neighboring countries in the planning process.

INTERNATIONAL ASSISTANCE

The proposal for the Action Plan was supported by the World Bank and UNEP.

MONITORING AND FOLLOW-UP

The proposed monitoring mechanism will determine whether allocated funds are used for the proposed actions and whether laws and regulations are consistent with actions proposed in the plan.

Ministries and government agencies will implement their portions of the Action Plan. At the state level, NEPA is responsible for monitoring and evaluating the actions of coordinating ministries and state agencies. In the State Council, the Environmental Protection Committee is in charge of evaluating and coordinating overall implementation.

OBSTACLES AND CHALLENGES

Conflicts did arise among different departments in preparing the Action Plan, particularly in the areas of fund allocation and administrative power. This problem has been solved, to a certain extent, through compromises arrived at by the Leading Group and/or Expert Team.

The frequent change of personnel during revisions of the plan was also a problem, causing, among other difficulties, inconsistencies from version to version. The differing opinions of the World Bank and its Chinese counterparts also caused confusion, which could have been dispelled if the Expert Team had met more frequently.

The lack of sufficient data and information, especially in the area of a national biodiversity database, also made it difficult to get a clear picture of the current status of ecosystems and species.

FACILITATING FACTORS

The support of participating government agencies proved very helpful to the preparation of the document. A draft Action Plan has been completed and sent out for review and comments.

COST, PERSONNEL, AND TIME REQUIREMENTS

It took about three years to complete the draft. The World Bank/UNDP's GEF invested about US$200,000. Approximately 150 experts participated in the process at different stages.

PLAN

The Biodiversity Conservation Action Plan was approved by the Environment Protection Committee of the State Council and was officially released at the China Biodiversity Action Plan Launching Conference held in Beijing in June, 1994.

10. Based on case study material prepared by Wang Sung, Executive Vice Chairman, Endangered Species Scientific Commission, Beijing, China; and Wang Enmin, Research Associate, Institute of Zoology, Chinese Academy of Sciences, Beijing, China.

■ *11. The Chilean National Biodiversity Action Plan*[11]

INSTITUTIONAL BASE

The National Commission on Environment (CONAMA) has been the lead government institution developing Chile's Country Study and National Biodiversity Action Plan. CONAMA was created in 1990 by presidential decree in response to the need to integrate environmental issues into the general political agenda. In March 1994, Congress approved the Framework Environmental Law, giving CONAMA a broad mandate to develop all environmental legislation and policies, advise the president on environmental issues, and coordinate the government's environmental activities.

CONAMA reports directly to the president of Chile through the ministry-level General Secretary of the Presidency. The General Secretary acts as the president of CONAMA's Council Directorate, which consists of representatives of 10 ministries: National Endowments, Economic, Development and Reconstruction, Agriculture, Transportation and Telecommunications, Health, Mining, Planning and Cooperation, Housing and Urban Development, and Public Works. The Ministries of Education and Foreign Affairs are also being incorporated into CONAMA's Council Directorate.

METHODOLOGY

In 1991, while the Biodiversity Convention was being prepared for the Rio Summit, CONAMA began coordinating the development of a biodiversity action plan based on its institutional mandate to seek equitable and sustainable use of natural resources within its strategy of economic development.

CONAMA initiated the process with a national assessment (country study) and ended it with a proposed Action Plan. During the assessment phase, CONAMA took a systemic view, which considers biodiversity to be both a cross-sectoral and multidisciplinary issue.

CONAMA invited a representative group of institutions to appoint an Expert Committee to the project. Eight professionals served on the committee, representing the public sector, the scientific community, the Ministries of Agriculture, Forestry and Fishery, CONAMA itself, the Museum of Natural History, the Science and Technology Council, the University of Chile, and the Forestry Action Plan of the Food and Agriculture Organization. CONAMA coordinated closely with various sectors and with the Ministry of Foreign Affairs, to integrate political and diplomatic issues with scientific and technological issues and to focus on both the international and national agendas.

After the design of a strategy that included the proposed Action Plan, CONAMA sponsored meetings in each of the 13 administrative regions of the country. It also sponsored a national survey of scientists and professionals from the public sector. The data compiled from these efforts were synthesized into a set of general conclusions. The proposed Action Plan was publicly launched, and comments were sought. After incorporating comments, CONAMA worked to incorporate biodiversity issues into the new government's Work Plan Agenda. In 1994, Chile ratified the Convention on Biodiversity. The National Assessment and the Action Plan were then updated to reflect the Convention.

CONAMA foresees continuing this process through a second and third period of consultation. Its 1994 Work Plan included newly prioritized biodiversity issues in the agenda of the Committee of Ministers:

- ratification of the Convention in Chile and follow-up of the Convention meetings (accomplished in September 1994)
- strengthening of CONAMA's role in the GEF international meetings (accomplished by April 1994)
- creation of a work plan and budget estimate for implementing the action plan (in process)
- establishment of a cross-sectoral Committee (in process), and
- preparation of a strategy for biodiversity legislation (in progress).

PARTICIPATION

The Expert Committee, consisting of professionals from major relevant institutions and constituents, was appointed to cost-effectively provide a cross-sectoral view, specific vision and expertise, and the consensus needed for the design of common goals. During the series of meetings in Chile's 13 administrative regions, more than 200 experts, scientists, and professionals were consulted.

The Expert Committee represented the following organizations: the Department of Natural Resources (DIPROREN) of the Agriculture Ministry; the Department of Natural Patrimony, charged with administering the National Protected Areas of the National Forestry Corporation; the Fishery Promotion Entity (IFOP); the Scientific and Technological Council (CONICYT); the Universidad de Chile; and the Museum of Natural History.

During the consultation stage, many other institutions, professionals, and scientists were involved. By the end, more than 500 consultations were held. Some 15 ministries and their institutes, 13 regional governments, 13 Regional Commissions of the Environment (COREMA), 25 provincial governments, 12 nongovernmental organizations, and the Council of University Rectories (formed by 22 universities throughout the country) were involved, along with other universities not associated with the Council. The Scientific and Technology Council, the Museum of Natural History, and the private sector contributed comments. Senators, Congress members, and representatives of both pro-government and opposition political parties also commented.

GOALS AND OBJECTIVES

The Action Plan was intended to establish the base for the conservation and sustainable use of Chilean biodiversity. These goals were determined through consensus of CONAMA's professionals, the Expert Committee, and the Executive Office.

RELATIONSHIP TO NATIONAL DEVELOPMENT PLANNING

Biodiversity was incorporated into the recently approved Framework Law of the Environment, which provides for establishing policy and regulations in all environmental matters, such as environmental impact assessment, preparation of norms, and management plans for natural resources. This ensures that biodiversity concerns will be incorporated into all national planning. CONAMA is also working to design and plan strategies that place biodiversity into integrated basin-management plans, soil-protection strategies, and climate change and ozone programs.

INTENDED TARGET OF PLANNING EFFORT

The reason for launching a separate document with both a synthesis of the Assessment and the Action Plan was to inform decision makers and the general public so that they can help implement the plan. The document also served to put the plan on the presidential agenda.

BIODIVERSITY CONVENTION

Because the Convention was still being prepared when Chile was drafting its strategy, the Assessment could not explicitly address the articles of the Convention. In 1992 and 1993, when the Convention was approved and being internationally ratified, a chapter was added in the Assessment to incorporate the specific articles of the Convention. These can easily be linked to the Action Plan. The Convention also provides international pressure and commitment, which speed up the process.

SCALE OF PLANNING EFFORT

The planning exercise addresses national-level ecosystems, species, and genetic materials. In some cases, these issues were analyzed at a local level. Chile did not directly involve neighboring countries in the planning process, nor did it consider transboundary issues in depth. At the global level, the Strategy addressed the need to consider effects of climate change and ozone depletion on biodiversity.

In the planning effort, biodiversity policies on legislation, education, scientific research, wildlife protection areas, river basins, fisheries, agroforestry and animal husbandry, and the Antarctic Territory were considered. So were ways to strengthen the public sector's capacity to address biodiversity issues.

INTERNATIONAL ASSISTANCE

CONAMA received financial support from the US Agency for International Development (USAID) under the guidance of the World Resources Institute during its first phase and during the development of the Biodiversity Strategy. A national consultant was hired to coordinate the initial phase. Technical support has been provided to team members by distributing IUCN, World Bank, WRI, UNDP, and UNEP documents and through participation in international workshops and seminars.

MONITORING AND FOLLOW-UP

A high-level cross-sectoral committee will be established to follow-up on the general process. The monitoring and follow-up of the plan will be undertaken by CONAMA, which will coordinate specific tasks with lead institutions. These institutions will implement actions related to their particular resource or activity and will present outlines of regularly scheduled reports to CONAMA. Twice a year, a general report will be presented to CONAMA's Committee of Ministers, and reports will also need to be presented to the financing institution.

OBSTACLES

The main problems were related to the novelty of the participatory approach and the discussion of biodiversity in the political arena. Other obstacles were initial feelings of territoriality and lack of communication between sectors. Lack of scientific documentation also delayed the process, as did the small initial budget.

FACILITATING FACTORS

CONAMA's recent creation and its environmental mandate facilitated the process.

CONAMA's involvement in other efforts, such as preparing the Framework Law on Environment, also benefited the process. In addition to helping prepare the Framework Law, CONAMA participated in education and public awareness campaigns; worked on issues of climatic change, ozone, and protection of air, water and soil resources; assisted in the drafting of the Forest Action Plan; and assisted in establishing a System for Environmental Impact Assessment.

Close coordination with the Foreign Affairs Ministries and the accreditation of technicians to the Intergovernmental Committee on the Convention on Biological Diversity (ICCBD) meetings and other related workshops allowed the introduction of international perspectives and technical advances into the national process. Cross-sectoral participation at the central, regional, and provincial levels also broadened the national discussion.

Submission of the Strategy to wide-ranging review greatly facilitated public discussion, the insertion of biodiversity in the political agenda, and the final ratification of the Convention.

COSTS, PERSONNEL, AND TIME

The process started in 1991 and is ongoing. The new administration has already included it in the work agenda of the ministries that form CONAMA.

Preparing and launching the proposed Action Plan took two years. The first phase of the project was financed by WRI and the US Agency for International Development (USAID) for US$10,000. The second phase has been financed by the Chilean government. The total estimated cost was US$30,000.

11. Based on case study material prepared by Consuelo Muñoz, Director, Biodiversity Program, National Commission on Environment (CONAMA), Santiago, Chile.

■ 12. *The Mexican Country Study on Biodiversity*[12]

INSTITUTIONAL BASE

In 1992, the president of Mexico created the National Commission for the Knowledge and Use of Biodiversity (CONABIO). CONABIO's fundamental task is to promote and coordinate the efforts of numerous Mexican institutions and groups along three lines: 1) knowledge of the country's biodiversity, through inventories, databases, and networking, 2) sustainable use, and 3) diffusion of knowledge about biodiversity to society.

The president of the republic is the president of CONABIO, and the minister of Social Development (SEDESOL) is its technical secretary. Nine ministers serve on the Commission which is staffed by a national coordinator and about 30 executive officers, analysts, and administrative personnel.

CONABIO coordinates the UNEP Country Study on Biodiversity for Mexico as part of the priority-setting process for the National Strategy on Biodiversity.

METHODOLOGY

CONABIO coordinates and edits the work of those directly involved with preparing the country study (as researchers, authorities, or participants) on each of the issues outlined in the Technical Annex of the UNEP *Guidelines for Country Studies on Biological Diversity*. The guidelines were not followed strictly but were adapted to Mexico's current capabilities and needs.

Mexico will be using only existing information from accessible sources, produced by institutions involved in constantly updating the information. This will allow the effectiveness of the Country Study's recommended measures to be monitored every 5 to 10 years.

The method adopted has the following steps:

- Analysis and synthesis of UNEP's *Guidelines for Country Studies*
- Adaptation of the document to Mexico's national needs and capabilities
- Identification of information sources, institutions, and experts on specific issues
- Invitation to external collaborators to produce texts, tables, cartography, and figures

- Information capture, synthesis, and editing by CONABIO
- Coverage of the following aspects by each of the sections of the study:
    - description of the information presented and evaluation of its reliability
    - diagnosis of the current situation and trends
    - suggested priorities for action, research agenda, and human resources needs
- Development of central text to give continuity to issues covered by the Country Study
- Ensuring that the country study has relationship to the articles of the Convention on Biodiversity and to specific national administration, research and social institutions and organizations
- Development of a draft for a biodiversity strategy to emerge from the Country Study; the strategy will be further evaluated and modified within the proposed planning process
- Probable production of a CDROM with text, raw data, and maps in accessible formats. Also, selected pieces of text and information will be made available through CONABIO's Internet GOPHER.

PARTICIPATION

Experts and responsible institutions were identified for each of the issues covered in the Country Study's table of contents. Participants are mostly from government or academic institutions, with a few nongovernmental national and international organizations (NGOs). Some participants are directly responsible for the administration of certain issues, whereas others are research groups.

The institutions that have already agreed to participate are: Instituto Nacional de Ecología, Secretaria de Desarrollo Social; Instituto de Geografía, Universidad Nacional Autónoma de México; Centro de Ecología, Universidad Nacional Autónoma de México; Centro de Investigación y Docencia en Economía; Instituto Nacional de Estadística, Geografia e Informática; World Wildlife Fund and Conservation International; and Székely & Associates (law firm).

Besides the groups mentioned, many other individuals, NGOs, and government agencies will participate.

GOALS AND OBJECTIVES

Mexico's Country Study will have the following sections:

- Social, Economic and Cultural
- Biodiversity
- Biological Resources
- Value of Biodiversity and Biological Resources
- National Capacity, and
- Draft for a Biodiversity Strategy.

The first three sections will contain the basic diagnosis of the situation. Section 4 is a baseline economic evaluation of biodiversity and biological resources. The "National Capacity" section emphasizes the institutional and governmental responsibilities, the assessment of research and human resource capabilities (including the indigenous and peasants point of view), and the identification of actors in the public and private sectors that should be involved in the strategy. Priorities for the Country Study are the compound result of evaluating both the current situation and Mexico's national capacity to confront the challenge. The synthesis of these five sections will allow CONABIO to establish the framework for a national planning effort with the capacity for long-term evaluation of the actions taken.

These goals and objectives were not established only for the Country Study process, but are within the functions assigned to CONABIO by presidential decree. However, the Country Study is giving Mexico a solid framework to use in evaluating the priorities adopted by CONABIO within a much wider scope.

RELATIONSHIP TO NATIONAL DEVELOPMENT PLANNING

The Country Study will provide the foundation for a National Biodiversity Plan. CONABIO's status as a high-level ministerial commission will allow it to promote and coordinate many of the actions identified within the final section of the Country Study. One aim of the study is to identify areas of public administration, research, and productive activities (with social and private sector involvement) that need to be reoriented, linked, and promoted. These findings will be translated into specific action plans in environmental, educational, legislative, investment, and commercial policy. The Salinas administration has a National Development

Plan (1988–1994) that did not specifically address biodiversity issues, although it did contemplate environmental, ecological, and sustainability components. It is likely that the next administration will specifically address biodiversity issues in its National Development Plan.

INTENDED TARGET OF PLANNING EFFORTS

The National Information System on Biodiversity (SNIB), now being developed, will be one of the most important products of CONABIO and is considered a user-oriented information service. The demand for information by the public, private, and social sectors is growing. Experts generating information (mainly taxonomical and ecological) are mainly in academic institutions. Their formats and objectives are not usually oriented to satisfy decision makers or public needs.

The information generated by CONABIO and external participants in the study will be put in formats that will eventually be accessible via the SNIB. The Country Study itself will contain summaries and syntheses, but many of the appendices will contain more extensive information, available in magnetic format, for decision making and public use.

RELATIONSHIP TO THE CONVENTION ON BIODIVERSITY

The Country Study design is closely related to the Biodiversity Convention. CONABIO has participated, along with the Foreign Affairs and Social Development Ministries, in following-up on the Convention. Also, a specific section of the Country Study addresses the legal and institutional issues involved in implementing the Biological Diversity Convention in Mexico.

SCALES OF PLANNING EFFORTS

The Country Study uses national and regional scales because a more local approach with national coverage was not feasible in the short term.

When the UNEP *Guidelines for Country Studies* were adapted to reflect national capabilities, the identification of shared resources with other countries was considered. CONABIO identified specific administrative and legislative instruments covering shared ecosystems, species, or genes. International collaboration will be minimal on this first country study, but, for example, an agreement was established to share mapping of vegetation and wetlands distribution with the World Wildlife Fund and Conservation International.

INTERNATIONAL ASSISTANCE

Resources for the Country Study were federally funded within CONABIO 's budget. Technical support was lent by UNEP and many other international agencies in the form of background materials.

MONITORING AND FOLLOW-UP

One of the main goals of the Country Study is for long-term follow-up to be performed easily. This explains the emphasis on raw data, both in tabular and cartographic form. The study will result in specific monitoring plans as part of the strategy. The implementation of the National Information System on Biodiversity has, among its purposes, to allow permanent collection and management of the information required for biodiversity planning and management.

OBSTACLES

A primary double-headed obstacle in preparing the Country Study is the acquisition of the basic information and its organization within a national framework. Mexico found that, like most nations, it does not have a solid methodological and theoretical framework for simultaneously analyzing social, economic, geographical, biological, productive, legislative, and administrative or political information. The development of such a framework is one of the most clearly identified tasks for the near future and will probably emerge as one of the study's action and research priorities. In Mexico's case, at least much of the information mentioned in the *Guidelines for Country Studies* exists (even if it is not immediately available.) But it is not easy to integrate, analyze, and make sense of as a solid planning base. Geographic Information System (GIS) technology will facilitate this integration and analysis.

Another problem is the lack of consistent and historical local information. Mexico has census information for the states dating from the beginning of this century. However, information at the municipal level is much harder to obtain because of changing boundaries and the differing categories under which information is gathered.

FACILITATING FACTORS

Academic institutions and the gradual growth of expertise in environmental and ecological public offices are factors that greatly facilitate the process of the Country Study. Another positive aspect has been the sensitivity of high- and middle-level policy-makers in Mexico's public administration. This receptiveness is not a casual change of mind in public officials but is the consequence of an international atmosphere of environmental and biodiversity awareness and of strong internal social pressure by rural and urban organizations, environmentalists, and the media.

COST, PERSONNEL, AND TIME REQUIREMENTS

CONABIO has assigned approximately US$60,000 to the Country Study. This amount is dedicated to information acquisition, salaries of auxiliaries, payment of some external contributions (when agency personnel can't help out), and editorial costs. However, this value does not account for all the full cost of the Country Study.

The Study is scheduled to take six months. Two people are assigned full-time in CONABIO, and at least 10 other members of CONABIO participate part-time. External collaborators are estimated at 50 individuals working in at least 20 different institutions.

12. Based on case study material prepared by Jorge Soberón, Executive Secretary, National Commission for the Knowledge and Use of Biodiversity (CONABIO), Jardines Del Pedregal, Mexico, and Jorge Larson, CONABIO, Jardines Del Pedregal, Mexico.

13. *The Polish Country Study on Biodiversity*[13]

Action for nature conservation has a long history in Poland, but there is no national strategy for the sustainable management of biodiversity. In the late 1980s, realizing the necessity for a national ecological development strategy, the Department of Environment Protection prepared a policy based on the principles of eco-development. This case study focuses on the preparation of a UNEP-sponsored country study on biological diversity. In Poland, it was the first summary to combine knowledge of biological diversity with attempts to assess its economic value. It constitutes starting material for a discussion on a future strategy and plan of action.

INSTITUTIONAL BASE

An agreement with UNEP headquarters in Nairobi put the Republic of Poland under an obligation to prepare a country study on the benefits, costs, and needs of protecting biological diversity. The Ministry of Environmental Protection, Natural Resources and Forestry contracted with the National Foundation for Environmental Protection (NFEP) to take on this task. The NFEP is among the largest and most dynamic NGOs in Poland, with solid experience in nature conservation, a staff representing a variety of fields, and good organizational and technical backup. The Foundation created a National Biodiversity Unit to prepare the Country Study.

METHODOLOGY

The study team restricted itself to existing data. The information system in Poland is not well-developed, so the hardest task was locating the necessary data. A pool of potential sources was created, and detailed questionnaires were prepared.

Economic data were generally compiled according to methods given in the UNEP *Guidelines for Country Studies*. However, the authors' lack of experience with these methods limited their use. As a result, in this section it was emphasized that the approach taken was far from perfect and accompanied by sizable errors. For example, the study concluded that Poland's financial need was considerably less than Germany's. This would seem to be unlikely given the realities of the new situation.

Since public awareness about biodiversity issues was low, sociological studies were conducted through public polls that focused on overall attitudes toward nature conservation and environmental protection. These were augmented by simulations and analyses by the press, and the evaluation of platforms of the different political parties.

PARTICIPATION

Selected specialist and scientific institutions were invited to help prepare the Country Study. The cost of involving state institutions is high compared with that of engaging private firms with greater "mobility" and expertise. The public is not participating heavily in the Study.

GOALS AND OBJECTIVES

The initial aim of the Country Study was to prepare the material UNEP needed to negotiate the text of the Convention. However, it soon became apparent that the longer-term aim was to prepare a cohesive policy for the protection and use of biological diversity, implementable at all possible levels of government. But decision makers do not yet understand that this task implies a long and laborious process.

RELATIONSHIP TO NATIONAL DEVELOPMENT PLANNING

The political and economic changes introduced in Poland four years ago have had many effects. The much-criticized centralized social and economic planning has been replaced by a market system. Although a Central Planning Office and the need for a national development plan still exist, the country has only now prepared a three-year plan. There are also no departmental development plans.

The Ecological Policy of the Country, passed by parliament in 1992, defines economic priorities in relation to environmental protection, as well as the necessary instruments of ecological policy. This document may be of decisive significance in the preparation of a plan for the sustainable development of the country.

INTENDED TARGET OF PLANNING EFFORTS

The planning process was supposed to draw on strategic documents at various levels of government. Discussion of the issues relating to biodiversity protection has also led those in scientific circles to draw up their own projects in this field and has encouraged environmental protection agencies to work out a method for monitoring of species.

Unfortunately, as already suggested, planning continuity and order are lacking. The documents prepared are also underutilized by the legislative and executive branches of government, the world of science and business, and social organizations and movements.

RELATIONSHIP TO THE CONVENTION ON BIOLOGICAL DIVERSITY

The process of ratifying the Convention is still at the preliminary stage, and no full analysis has yet been made of Poland's obligations under the Convention. The Ministry of Environmental Protection, Natural Resources and Forestry has approached the Institute of Environmental Protection to define the legal, scientific, and financial implications of implementing the Convention and to set out the consecutive tasks in this area. All resolutions of the Convention must find a place in any future strategy for biodiversity protection.

SCALES OF PLANNING EFFORTS

The preparation of the Country Study has clarified the need for a national biodiversity strategy, but the scope of such a strategy is so far difficult to define. The strategy will certainly find points of reference on the international and national levels, as well as at the regional, provincial *(voivodeship)*, and community *(gmina)* levels.

INTERNATIONAL ASSISTANCE

The Polish Country Study was prepared with the economic and technical assistance of several international organizations. The Study was prepared at the request of UNEP, in accordance with strict guidelines. Although the guidelines set out specific procedures for document preparation, organizations were free to choose the methods to be applied. In the course of preparing the case study, consultants acting on behalf of UNEP offered their comments.

Materials sent by the World Conservation Monitoring Centre, among others, were also used in preparing the study. Unfortunately, the short time available made it impossible to compare the study team's approach with that of an analogous study in Germany. Such cooperation between countries of similar status would be of exceptional value if time and circumstances allow.

MONITORING AND FOLLOW-UP

No formal process currently exists for monitoring biodiversity in Poland, although work has begun to create appropriate databases and a program for monitoring living natural resources. The Polish Parliament's ratification of the Convention will also result in work on the adaptation of various regulations and programs.

A national Country Study prepared at regular intervals (for example, every five years) could constitute an ideal method for monitoring changes in biodiversity at the genetic, species, and ecosystem levels, as well as the costs, benefits, and needs related to conservation. Such a study would also help define research priorities and scope.

OBSTACLES

The main obstacle to preparing a biodiversity conservation program is the lack of uniformity and consistency in the actions of responsible parties, which is the result of frequent political changes.

Poland's weak information technology is also a serious hindrance. In the course of the Country Study it became obvious that various sectors and organizations lacked knowledge about work being conducted elsewhere in the country. Experience illuminated cases where some projects lacked support by key groups while others faced overlapping mandates and implementing activities.

Documentation is inexpensive and it can be funded through national or foreign sources. However, projects have been put into effect to a very limited extent. For instance, financial constraints have led to the closure of both seed banks for old varieties of crops and collections of livestock that are unique in the world.

FACILITATING FACTORS

There is no doubt that the composition of the study team made the preparation of the Country Study easier. Indeed, it was vital for gaining access to certain materials and was of particular significance with such a short time allotment.

COSTS, PERSONNEL, AND TIME REQUIREMENTS

In effect, three months were available for the preparation of the Country Study. As a result, a relatively small team was used, and it was extended by consultation where necessary. The total cost of the Country Study, US$65,000, was covered in its entirety by UNEP. Approximately 70 people took part in the work, including the core team of some 20 people.

13. Based on case study material prepared by Andrzej Weigle, Ecological Studies Promotion Office, National Foundation for Environmental Protection, Warsaw, Poland.

■ *14. The Kenyan Country Study on Biodiversity*[14]

INSTITUTIONAL BASE

The National Museum of Kenya (NMK) is the national repository of all of the country's prehistorical, cultural, and biological heritage. It carries out basic and applied research on all disciplines under its custodianship and disseminates this information through educational programs and public exhibits. The newly established Centre for Biodiversity is specifically set up to address biodiversity issues and strengthen scientific collaboration with both national and international organizations. In 1991, the Kenyan Government gave the NMK the mandate to establish the Centre for Biodiversity and coordinate all biodiversity work in the country.

Between November 1991 and April 1992, the government, through the National Biodiversity Unit (NBU) based at the NMK and in cooperation with UNEP and the Overseas Development Administration (ODA), undertook a biodiversity country study on the cost, benefits and unmet needs of biological diversity conservation in Kenya. The National Environment Secretariat (NES) also took a lead in this study.

NES is a government agency under the Ministry of Environment and Natural Resources. Although NES has broader responsibilities on matters relating to the environment than any other national organization, it lacks the personnel needed to coordinate environmental issues. The Kenyan government, with GEF support, has appointed a consultant to address the role of NES.

METHODOLOGY

A team of experts who would outline the Country Study was established by the Governing Council of UNEP. The National Biodiversity Unit then prepared the terms of reference and objectives, while the NMK provided most of the specialists. Other institutions and individuals were also involved. The Country Study is now being used as a reference document in the preparation of the National Environmental Action Plan (NEAP). The report assesses the nature and extent of Kenya's biodiversity and the way it has changed in recent years. Reasons for these changes, the measures and costs of conserving the country's biological diversity are discussed in the report.

PARTICIPATION

The Country Study on biodiversity was coordinated and conducted by a team of Kenyan experts drawn from 13 government organizations concerned with biodiversity conservation, three universities, and three relevant NGOs. All were chosen for their expertise and involvement with local communities. ODA provided additional technical assistance.

The NEAP (in progress) is attempting to provide a framework for integrating environmental considerations into the nation's overall economic development. Several government parastatal organizations, universities, nongovernmental organizations, and individuals have been involved in the preparation of the NEAP document.

GOALS AND OBJECTIVES

The goal of the Kenyan Country Study was to help the UNEP Governing Council and Inter-Governmental Negotiating Committee arrive at a more accurate and realistic assessment of the total benefits, costs, and needs of conserving and sustainably using biodiversity and biological resources.

The operational objectives of the Country Study are to identify and acquire the current knowledge about biological diversity in Kenya, determine the costs of meeting information needs, assess current pressures on biological resources and future trends, assess the present and potential value to humankind of Kenya's biodiversity, establish priorities for conserving this biodiversity, and develop methods of implementation.

RELATIONSHIP TO NATIONAL DEVELOPMENT PLANNING

In 1963, Kenya opted for an economic and social development strategy that could guide the more efficient and cost-effective use of the natural, human, and financial resources at the country's disposal. This planning process has evolved into long-term prospective planning, national five-year development plans, sessional papers, district plans, various sector-based plans, and working parties. Within the government structure, coordination of the planning process is currently under the Ministry of Planning and National Development, which will coordinate the Country Study on Biodiversity.

Other related plans that have been or are being prepared include the Forest Master Plan, the Kenya Indigenous Forest Conservation Profile documents, the National Environmental Action Plan, the UNCED National Report for Kenya, and the Environmental Policy for Sustainable Development (National Development Plan).

INTENDED TARGET OF PLANNING EFFORTS

The target groups for the Country Study were government policy-makers, politicians, local leaders, and heads of resource departments.

RELATIONSHIP TO THE CONVENTION ON BIODIVERSITY

The Country Study explicitly addressed the articles of the Convention. The current preparation of the NEAP is a step toward developing the national plan for conservation and sustainable use of Kenya's biological diversity.

SCALES OF PLANNING EFFORTS

The biodiversity-planning exercise addressed regional, local, and national issues. It considered ecosystems, species, and genetic material, whether shared, or related to regional or other conventions and international agreements. The Country Study did not attempt to categorize which genetic materials are shared. There was no cooperation with neighboring countries in the planning process, although Uganda was also conducting a Country Study.

INTERNATIONAL ASSISTANCE

Support for the Country Study came from the British ODA. The NEAP Process receives support from the World Bank. The British Government funded the study through bilateral agreement.

MONITORING AND FOLLOW-UP

No concrete monitoring process was proposed, but follow-up activities were identified. In the process of implementing the Country Study, the experts involved were assigned specific responsibilities. However, no responsibilities for implementing identified activities were assigned to institutions or individuals. Everything was left in the hands of the NBU. Due to lack of financial support, the NBU cannot implement proposed activities; these are being addressed and implemented by the Kenyan government (e.g., NEAP, the Forest Master Plan).

OBSTACLES

A lack of institutional coordination hindered the process, as did conflicts within the lead agency for biodiversity and the absence of an inventory of scientific institutions dealing with relevant biodiversity areas. The process also suffered from the lack of a research information database and personnel trained in biodiversity assessment. The policy framework and the political commitment to the process were weak, and because biodiversity expenditure is not a "pure" category in government budgets, it is difficult to evaluate and track these expenditures. Kenya also has no policy on land use, and it lacks clear laws on the use of natural resources.

Most of the institutional obstacles were overcome by forming the NBU and drawing resource people from different institutions. Scientific and other legal or policy obstacles were identified, and recommendations to overcome them were addressed.

FACILITATING FACTORS

The availability of funds and the government's commitment to the conservation of biological diversity encouraged the participants, who tackled their assignments effectively and successfully. Key government institutions, especially the NMK, also assisted with basic data and documents.

COSTS, PERSONNEL, AND TIME REQUIREMENTS

The Country Study took four months, whereas NEAP has already taken five months. The Country Study cost approximately US$258,000. At the request of UNEP, the British government financed the process through bilateral agreement. Approximately 70 people worked on the Country Study.

14. Based on material prepared by Mohamed Isahakia, Director/Chief Executive, National Museums of Kenya, Nairobi, Kenya; and Richard Bagine, Centre for Biodiversity, National Museums of Kenya, Nairobi, Kenya.

■ *15. The Egyptian Country Study on Biodiversity*[15]

INSTITUTIONAL BASE

The Egyptian Environmental Affairs Agency (EEAA) was established in 1982 as an affiliate of the Council of Ministers. The EEAA sets and implements national environmental policies, including natural heritage conservation. It is supported by a national program for environmental research and studies sponsored by the Academy of Scientific Research & Technology, and several nongovernmental organizations (which include women and youth), in addition to all national agencies concerned with environmental affairs.

In June 1992, Egypt signed the Convention on Biological Diversity. Immediately afterward, a core National Biodiversity Unit (NBU) was established under the EEAA, Department of Natural Protectorates, to prepare both a Country Study on the status, costs, benefits, and unmet needs of biodiversity conservation in Egypt and a national biodiversity strategy. The Country Study is currently in progress, and the National Biodiversity Strategy is in draft.

METHODOLOGY

The Country Study on Biodiversity followed the methodology in the UNEP Guidelines for Country Studies. By the end of 1994, the Country Study was expected to contribute to or determine:

- baseline information on biodiversity in Egypt;
- biological data on species, habitats, and ecosystems and its *ex situ* and *in situ* management;
- defined priority areas and programs for effective conservation of biological diversity in Egypt;
- the costs of biodiversity conservation and its rational use in Egypt;
- data on the economic values of species, biological resources, and ecosystem services;

- the training of national personnel in assessing biodiversity;
- a national biodiversity monitoring unit that could developed into a monitoring center; and
- a national biodiversity strategy and action plan as a follow-up to the Country Study.

NBU experts are now gathering data on biodiversity status and preparing reports on all aspects of biodiversity. The Country Study includes identification of habitats, sites, species, and genomes of national importance; identification and ranking of problems within Egypt; and identification of conservation measures and areas of significant biodiversity.

PARTICIPATION

Members of the NBU are experienced staff representing different sectors of government organizations, including universities, agricultural research centers, the National Institute of Oceanography and Fisheries, the National Planning Institute, the National Research Center, the Ministries of Agriculture, Interior, Scientific Research, Irrigation and Water Resources, and Tourism, the Egyptian Wildlife Service, Giza Zoological Gardens, and the Academy of Scientific Research and Technology. NGOs include the Society for the Preservation of Natural Beauty, the Egyptian Society for Preserving Natural Resources, the Egyptian Society for Landscaping, the Friends of the Trees Society, and the Friends of Marine Life Society.

GOALS AND OBJECTIVES

The objectives of Egypt's Country Study are to assess the status of national biodiversity and identify gaps in knowledge, basic needs for effective conservation, and rational use of biodiversity; to outline the support measures and costs of meeting those needs; and to evaluate the benefits of implementing these measures.

RELATIONSHIP TO NATIONAL DEVELOPMENT PLANNING

In 1992, 10 working groups of national experts, drawn from various ministries, institutions, and nongovernmental organizations, were brought together to draw up a high priority National Environmental Action Plan (NEAP) for the country. The initial outcome was presented to, and thoroughly discussed with, a widely representative and highly qualified forum. The output of this forum was further scrutinized by a team led by the World Bank and made up of representatives from Canada, Denmark, the European Community, Italy, Netherlands, Norway, the UK, UNEP, the United Nations Development Programme (UNDP), and the U.S.

INTENDED TARGET OF PLANNING EFFORTS

Development of the NEAP, Country Study, and National Biodiversity Strategy included the participation of government policy-makers and other decision makers. All three efforts are targeted at providing policy and decision makers with information and options necessary for management decisions.

RELATIONSHIP TO THE CONVENTION ON BIODIVERSITY

The Country Study benefits from the technical assistance of UNEP and should facilitate Egypt's implementation of the articles of the Convention. Although the Country Study is not intended to address the articles of the Convention specifically, it provides a foundation from which to develop a National Biodiversity Strategy.

SCALES OF PLANNING EFFORTS

Egypt is a signatory of many global conservation treaties, including the African Convention on the Conservation of Nature and Natural Resources (African Convention, 1968), which is considered the most comprehensive multi-lateral treaty for the conservation of nature. It requires parties to establish conservation areas for ecosystem protection and scientific conservation plans for protecting other important resources.

INTERNATIONAL ASSISTANCE

Technical and/or financial assistance for the preparation of the National Environmental Action Plan and the Country Study was provided by the World Bank, UNEP, UNDP, and experts from Canada, Denmark, Italy, Netherlands, Norway, the UK, and the U.S.

FOLLOW-UP

The implementation of the proposed national strategy will consist of two phases. Actions scheduled in the next two years are: the cataloging of existing knowledge, the initiation of scientific planning, the securing of financial support, the engagement of NGO participation, the planning of field research, the initiation of conservation measures, the engagement of international cooperation, and the researching of legal and institutional concerns. Actions scheduled during the next 20 years are: continued documentation of ongoing research, scientific research on monitoring and surveying, the expansion of financial support, continued NGO participation, field research, the implementation of conservation measures, continued international cooperation, and the development of legislation to strengthen the efforts undertaken as part of the National Biodiversity Strategy.

15. Based on case study material prepared by Esam Ahmed Elbadry, Head of the Biodiversity Unit, Department of Natural Protectorates, Egyptian Environment Affairs Agency, Cairo, Egypt.

■ 16. *The Costa Rican Country Study on Biodiversity*[16]

INSTITUTIONAL BASE

Well before the negotiations for the Convention on Biodiversity, Costa Rica planned a series of measures to conserve biodiversity. In 1988, the National Conservation Strategy for Costa Rican Sustainable Development (ECODES) and the Forestry Action Plan (PAF) were completed. In 1992, the Country Study on Biodiversity was prepared, and in 1993, the first phase of the Strategy for the System of National Conservation Areas (SINAC) was completed. The Ministry for Natural Resources, Energy and Mines, created in 1986, was the lead agency for all these studies. The University of Costa Rica, the National Museum, and the National Biodiversity Institute (INBio) also helped create a *de facto* national biodiversity strategy by integrating policies and goals, and a long-term vision into other plans.

METHODOLOGY

Each of the above-listed studies and processes used specific methodologies, with the participation of all levels of society that are directly involved in this field.

ECODES involved more than 150 professionals. An executive secretary was appointed to establish administrative mechanisms and to direct and supervise the strategy's development. Nineteen groups coordinated the sectoral strategies with a support group for each sector. Two advisory committees, a director, and other technical support were provided to the secretary. Three inter-sectoral meetings and two workshops, including a National Congress where the strategy was presented, allowed for dynamic participation in reaching a consensus on environment and development problems.

The PAF process followed the methodology proposed by the UN Food and Agriculture Organization (FAO). The Country Study followed the outlines put forth by UNEP, with methodological adaptations and improvements in the presentation and analysis of information. A National Biodiversity Unit (NBU) was formed to supervise and approve the study. INBio, MIRENEM, and the National Museum formed an Inter-institutional Technical Committee. The work concluded with a National Consultation involving both private and public institutions.

The SINAC strategy was an internal planning process undertaken by the Planning Department of the National Parks Service, with administrative, evaluation, and personnel support from the University of Costa Rica. The National Parks Service provided technical and financial analysis of the management and capacity needs for biodiversity within the protected areas under the new concept of decentralized administration and the local community involvement.

PARTICIPATION

Public institutions, NGOs, the private sector, scientists, politicians, and other interested parties participated in the biodiversity-planning process. After the documents were drafted, participants assisted in executing specific biodiversity conservation activities and in new processes, such as planning a total biodiversity inventory in the Guanacaste Conservation Area (120,000 ha), INBITTA. This effort was coordinated by INBio and represents the first effort of its kind to inventory all the taxa in a protected area. The project is steered by an advisory committee of more than 20 public and private institutions.

GOALS AND OBJECTIVES

Each of the studies put forth its own objectives for the conservation of biodiversity. These objectives include:

- to save the rich biodiversity present in the country, by consolidating the system of national conservation areas;
- to know what is present by means of a national biodiversity inventory in which information is compiled and arranged in a usable way; and
- to investigate and develop technology for the non destructive use of biodiversity.

RELATIONSHIP TO NATIONAL DEVELOPMENT PLANNING

Several of Costa Rica's studies, strategies, and plans incorporate a biodiversity component. In 1993, Costa Rica also revised its forest sector planning with the technical assistance of the World Bank. The Costa Rican biodiversity strategy is not synthesized in a formal document, however. Instead, the principles are contained in different sectors of management planning.

INTENDED TARGET OF PLANNING EFFORTS

The flexible political structure has allowed decision making within the framework of the *de facto* National Biodiversity Strategy. Examples of these decisions are the creation of INBio and the evolution of the national protected areas into a landscape-scale system of "conservation areas."

RELATIONSHIP TO THE CONVENTION ON BIODIVERSITY

Before the Convention on Biodiversity was ratified in June 1994, Costa Rica had already addressed many of its recommended actions. Given the creation of the position of secretary of sustainable development and with the concern expressed by the government, INBio, and the secretary, can now incorporate the principles established in the Convention into all of Costa Rica's plans and strategies and will.

SCALES OF PLANNING EFFORTS

Costa Rica has been active on all three levels of biodiversity management: local, national, and regional. Through the Central American Commission on Environment and Development (CCAD), the Central American presidents approved the Central American Biodiversity Treaty. In 1993, INBio held a workshop for Latin American and Caribbean countries on institutions for the management of biodiversity.

INTERNATIONAL ASSISTANCE

The Country Study was financed by the Canadian government; ECODES was financed by an international group of NGOs, including IUCN, WWF-USA, Conservation International, the Conservation Foundation, The Nature Conservancy, the University of Costa Rica, and others. Numerous NGOs, foreign governments, and international organizations also played roles in the implementation process.

MONITORING AND FOLLOW-UP

Costa Rica has not implemented a permanent mechanism for monitoring and follow-up of biodiversity activities. MIRENEM with INBio, by means of the Convention, is responsible for the biodiversity inventory in the national conservation areas as well as the biodiversity prospecting activities. The Convention has allowed for mutual sharing of responsibilities among institutions.

OBSTACLES

In conceptualizing the plan, the principal obstacle was indecision about the scope of SINAC, and, occasionally, lack of continuity resulted from changes in key functions within MIRENEM. This situation has caused other problems, limiting inter-institutional coordination and access to information.

FACILITATING FACTORS

Facilitating factors for this process include the presence of highly qualified personnel, appropriate political decisions, and legislation that does not hinder the planning process. The most important factor has probably been the tradition in Costa Rica of facilitating and encouraging innovation.

The plans have been approved by executive power, and their implementation has become the duty of the public and private sectors.

COSTS, PERSONNEL, AND TIME REQUIREMENTS

An estimate cannot be made of the costs of specific biodiversity-planning efforts.

16. Based on case study material prepared by Rodrigo Gámez, Director, National Biodiversity Institute (INBio), San Jose, Costa Rica.

■ 17. The South Pacific Biodiversity Conservation Program[17]

The South Pacific Biodiversity Conservation Programme (SPBCP), a coordinated regional program and strategy to establish a system of conservation areas, is the centerpiece of biodiversity planning in the South Pacific region.[18] No country in the South Pacific has prepared a national biodiversity strategy. However, several countries are preparing plans for establishing biodiversity conservation areas. The South Pacific Biodiversity Conservation Programme seeks to coordinate those efforts.

INSTITUTIONAL BASE

Task forces from agencies such as Environment, Forestry, or Natural Resources in collaboration with other agencies and NGOs were established at the suggestion of the South Pacific Regional Environment Programme (SPREP) to prepare national conservation area plans.

Studies undertaken by outside institutions such as universities or by donor-funded projects in forestry resulted in some site-specific biodiversity plans. Regional and global programs such as the GEF, the Tropical Forestry Action Plan, the Biodiversity Support Programme, and bilateral aid programs served as stimuli for other countries to prepare similar plans.

METHODOLOGY

In many countries of the South Pacific, biodiversity conservation has been possible in national parks and reserves on government land without support from local communities. However, the areas suffered degradation as people encroached and bordering societies took up incompatible activities.

Future conservation efforts will call for governments and other interested parties to gain access to unpreserved land and to work in partnership with landowning communities to protect the biodiversity on their land. The conservation areas will be managed jointly with the local communities and will be owned by them. Coordinating groups comprising representatives from the landowning communities, appropriate government agencies, NGOs, and other interested parties will plan and implement activities in the areas.

The procedure followed in the planning for conservation areas under the SPBCP is as follows: the conservation-area concept was introduced by SPREP and accepted by regional governments and NGOs. Countries prepared, and submitted for SPBCP consideration, conservation-area-concept proposals based on the SPBCP philosophy. Conservation-area-concept proposals are approved by the Technical and Management Advisory Group (TMAG) of the SPBCP. Long-term project plans are prepared by SPBCP in collaboration with task forces. Long-term plans are approved by TMAG. Plans are implemented by governments, NGOs, landowners, and SPBCP.

PARTICIPATION

At the regional level, government representatives, regional NGOs, and institutions were involved in the preparation of the design of the SPBCP. At the national level, the task forces and teams—which include representatives of government agencies, NGOs, and the private sector—were involved in the preparation of the concept and project plans for specific conservation areas.

Government agencies commonly represented include: the Environment and Conservation Agency (usually the lead agency), Forestry, Agriculture, Tourism, Central Planning, Public Works, Education, Health, Fisheries, and Village Affairs. Nongovernmental organizations include: Ministries for Women and Youth Affairs, environmental or conservation NGOs, local and village groups, regional universities, and research institutions.

GOALS AND OBJECTIVES

The objective of the South Pacific Biodiversity Conservation Programme is to protect the biodiversity of the region through the establishment of conservation areas.

RELATIONSHIP TO NATIONAL DEVELOPMENT PLANNING

For many countries, the conservation area plans were developed independently of national plans. However, 15 countries of the South Pacific have either completed or are in the process of completing National Environmental Management Strategies (NEMS), which aim to integrate environment and development.

In several NEMS, high-priority biodiversity projects have been identified, and these have been recommended for implementation under the SPBCP.

Groups of task forces prepared State-of-the-Environment reports, the reports to UNCED, the NEMS, and the Biodiversity Conservation Area Concept Plans which contributed to continuity throughout the documents.

At the regional level, the Action Plan for the Protection of the Environment of the South Pacific (SPREP Action Plan) and the Action Strategy for the Conservation of Nature in the South Pacific (a joint initiative of SPREP and IUCN) provide the blueprints for action. Both these plans are reviewed and revised every five years to accommodate new changes in national and regional priorities, and they are executed by SPREP.

INTENDED TARGET OF PLANNING EFFORTS

During the formulation of the SPBCP, a draft project document was circulated to all eligible countries for their review and approval as necessary. UNDP required that at least seven countries should indicate their acceptance of the document by signing it.

The SPBCP final project document was adopted by all member states of the South Pacific during the SPREP annual meeting in 1992, after review and acceptance by several technical working groups of Pacific Island experts.

RELATIONSHIP TO THE CONVENTION ON BIODIVERSITY

The planning for the SPBCP was completed before the negotiations on the Global Biodiversity Convention were completed. However, specific attention was given to the important linkages between the Programme and the Biodiversity Convention. Provisions of Articles 6 and 8 of the Convention were cited in the project document as providing guidance for the regional program. The SPBCP will work within the guidelines and spirit of the Convention and will encourage other countries participating in the Programme to become signatories to the Convention. It remains to be seen how the individual countries themselves will more effectively link their conservation area plans with the Convention.

SCALES OF PLANNING EFFORTS

Although the SPBCP promotes closer cooperation between countries in the conservation of the biodiversity of the region, its program activities are largely site-specific. To compensate for this, the Programme will provide additional funds to the SPREP to help address other biodiversity programs relating to the protection of marine mammals, marine turtles, and avifauna. The SPREP will also be responsible for the coordination of country activities under other regional and international conventions such as the CITES, World Heritage, and Ramsar Conventions.

INTERNATIONAL ASSISTANCE

Funding for the SPBCP came from the GEF, with co-financing by the government of Australia. Ten million dollars were made available for biodiversity conservation in 14 countries of the region over a five-year period.

MONITORING AND FOLLOW-UP

Evaluation and follow-up procedures for both the SPBCP and individual conservation area projects were proposed in the project document for the SPBCP in accordance with UNDP procedures. Annual and midterm reviews will be undertaken jointly by UNDP and SPREP, and tripartite reviews will be held jointly with the participating countries.

OBSTACLES

Some of the problems faced by island countries in the preparation of biodiversity plans include the following: lack of trained people who could lead efforts in the preparation of the plans, lack of coordination between governments and local NGOs, and complex customary tenure regimes in the Pacific which often necessitate long and time-consuming negotiations for securing lands for biodiversity conservation programs.

FACILITATING FACTORS

The creation of the SPBCP and specific conservation-area projects within the countries were facilitated by the availability of GEF resources. The SPBCP, by promoting "locally owned and locally managed" conservation areas, was also a facilitating factor.

COST, PERSONNEL, AND TIME REQUIREMENTS

The SPBCP took almost two years from the time the concept paper was first drafted to the approval of the final draft project design at a cost of about $850,000. This money came from the GEF through UNDP.

One person was involved full-time for six weeks with the preparation of the concept document; one was involved part time. One person was involved full-time for three months with the preparation of the Project Formulation Framework (PFF). One person was involved full-time for eight months with the preparation of the project plan; three others were involved part-time. Including the reviewers, government and NGO officials at SPBCP meetings, and secretariat staff, the estimated number of people involved in the preparation of the plans is about 500.

17. Based on case study material prepared by Iosefatu Reti, Program Manager, South Pacific Regional Environment Program, Apia, Western Samoa.

18. Countries eligible for funding under the Programme include Cook Islands, Federated States of Micronesia, Fiji, Kiribati, Marshall Islands, Nauru, Niue, Palau, Solomon Islands, Tokelau, Tonga, Tuvalu, Vanuatu, and Western Samoa.

Annex B • Background Information on Intergovernmental Decisions and Corresponding Activities that Provide Overall Mandate for this Study

Background Information on Intergovernmental Decisions and Corresponding Activities that Provide Overall Mandate for this Study[1]

1. Following decisions of UNEP's Governing Council, the negotiations for a Convention on Biological Diversity were successfully concluded. It was opened for Signature at the Earth Summit in Rio de Janeiro on 5 June 1992. By 13 September 1994, the European Community and 176 States had signed the Convention, and 89 States had ratified it. The Convention entered into force on 29 December 1993, and the first meeting of the Conference of the Parties took place in Nassau, The Bahamas from 28 November to 9 December, 1994.

2. At the close of negotiations, governments believed so strongly in the urgency of reversing the worldwide loss of biological diversity that they adopted three resolutions calling for immediate action. Unwilling to wait until the Convention enters into force, governments defined a wide range of issues for further intensive study to ensure that the momentum established during the negotiation phase was not lost. The Conference for the Adoption of the Agreed Text of the Convention on Biological Diversity which took place in Nairobi on 22 May 1992, recognized the need to take immediate action for the conservation and sustainable use of biological diversity, pending entry into force of the Convention, and adopted several resolutions calling for various types of action during the interim period. The Nairobi Final Act conveyed the resolutions and the agreed Text of the Convention to the Rio Fourth Summit. Resolution 2 of the Final Act of the Conference for the Adoption of the Agreed Text of the Convention on Biological Diversity, in particular, set out a challenging agenda that was pursued by the Intergovernmental Committee on the Convention on Biological Diversity (ICCBD) in preparation for the first meeting of the Conference of the Parties.

3. The provisions of the Convention and the thrust of Resolution 2 of the Nairobi Final Act are reinforced by *Agenda 21*—which, in Chapter 15, stresses the value of biological resources as a capital asset with great potential for yielding sustainable benefits at the country level. It highlights and emphasizes the need to build capacity for the assessment, study, evaluation and monitoring of biodiversity at the national level, whilst ensuring the full participation and support of local communities. It also calls for the production of Country Studies, with particular reference to costs, benefits and socio-economic issues relevant to effective biodiversity conservation and sustainable use of biological resources. Article 6 of the Convention calls upon the governments to develop national strategies, plans or programs for the conservation and sustainable use of biological diversity, or adapt current strategies, plans or programs for this purpose.

4. In discussing global biodiversity conservation needs and costs during the negotiation of the Convention, the *Ad Hoc* Working Group of Experts on Biological Diversity established by the Governing Council of UNEP to consider within a broad socio-economic context the technical content of the Convention, identified the need to carry out an in-depth study on the cost of basic conservation needs, taking into consideration the level of funding currently provided by existing development and other assistance programmes for the purpose of conserving biodiversity in developing countries. Initial cost estimates suggested one to ten billion dollars would be needed yearly over the next 10 to 15 years to meet priority conservation needs identified by the Working Group.

5. To sharpen the estimates of projected costs, the Technical Group, as well as the Intergovernmental Negotiating Committee for a Convention on Biological Diversity, established by the Governing Council with a mandate to negotiate a global Convention on Biological Diversity, recommended the preparation of country specific case studies, taking into consideration the full range of biodiversity and ecosystems, and the benefits generated as well as costs incurred in investing in biodiversity conservation. It is expected that these studies will be comprehensive, providing data not merely on the order of magnitude of costs but also on the status of biodiversity and the way in which biological diversity is estimated and valued.

6. The *Ad Hoc* Working Group on Biodiversity of the Scientific and Technical Advisory Panel (STAP) of the Global Environment Facility (GEF) pointed out at its meeting in Washington D.C. in June 1991, the importance of carrying out the Country Studies. Its opinion is that, ideally, such studies should be done in all countries, and investments in biodiversity were considered of limited value without them.

7. To facilitate harmonization and comparability of the results obtained from different countries:
   - guidelines, including methodology and format for the preparation of these studies were prepared, distributed to experts and governments for comments and advice, and reviewed and finalized by a multidisciplinary team of experts hosted by the Government of Canada in Montreal in April 1991.
   - an international Steering Committee and Advisory Team (including biologists, resource economists, data analysis and management experts) were established by the executive director of UNEP to provide advice and guidance to countries involved in the preparation of the studies and to ensure common understanding and uniformity in the application of the Guidelines.
   - the World Conservation Monitoring Centre (WCMC) in Cambridge, U.K., advised individual countries in the preparation of the studies through, among other things, the transfer of biodiversity data, provision of relevant bibliographic material, review of appropriate information held by other organizations outside the country, provision of expertise in the field of biodiversity information analysis and management, and participation in Advisory Team missions to the countries.

8. At the country level, National Biodiversity Units (NBUs) were established to coordinate and oversee the preparation of the country studies. The composition of the NBU was multi-sectoral/multidisciplinary covering a wide range of institutions and expertise: government departments, national universities, national museums, non-governmental organizations, wildlife agencies, private sector, and national as well as external consultants. It should be noted that the country study exercise was primarily a data-gathering exercise and not a field survey.

9. The pace of ratification of the Convention and the preparation of biodiversity country studies, strategies, and action plans has been increasing steadily. The primary objective of the country studies initiated by UNEP in 1991 is to assist national governments to identify, in the light of social, economic, environmental and other objectives, the basic needs and levels for effective conservation, including rational use of national biological resources and the necessary supportive measures and costs to meet those needs, as well as the benefits associated with the implementation of these measures. The country studies are also expected to: (a) provide an overview of the status of biological diversity, in terms of present knowledge of conservation efforts and future conservation needs and costs; and (b) institutionalize national biodiversity conservation strategies and action plans to be carried out in concert with national, regional and international institutions, and within the framework of the Convention on Biological Diversity and *Agenda 21*.

10. Using UNEP Guidelines and supported by professional backstopping through an *ad hoc* Expert Advisory Team for Country Studies established by UNEP, the first tranche of country studies was completed by May 1992, including studies from 10 countries, namely, Bahamas, Canada, Costa Rica, Germany, Indonesia, Kenya, Nigeria, Poland, Thailand, and Uganda. A Synthesis Report on the results from these ten countries submitted by January 1992 (UNEP Country Studies/Inf.1, Nairobi 23 April 1992) was prepared by UNEP.

11. At its final meeting in February 1992, the Advisory Team recommended that the Guidelines for Country Studies issued in May 1991 (UNEP/Bio.Div./Guidelines, May 1991) should be revised to incorporate the experience gained from the first tranche of studies. In particular, the revised guidelines should focus more on the compilation of biological and economic data to reinforce biodiversity planning within countries. Coincidentally, they should bring to the fore a basis for quantifying national unmet needs for funding to implement the Convention on Biological Diversity, which had been the primary orientation of the first tranche of studies.

12. A thorough and systematic revision of the 1991 guidelines entailed the establishment of four task forces to review, revisit, and develop comprehensive guidelines for national assessments of biodiversity status and trends, as well as accurate reflection of economic costs incurred and benefits derived from the conservation and sustainable use of biodiversity. Each task force produced a draft report for consideration and review by the Costa Rica Conference convened for this purpose, among others. A new *Guidelines on Country Studies for Biological Diversity* was produced by UNEP in May 1993, and is being used by a number of countries for the preparation of studies.

13. The May 1993 guidelines relate principally to the preparation of Country Studies. This present set of guidelines relate to National Strategies (Articles 6, 10, 11, 12, 13, and 14) and Action Plans (6, 7, 8, 9, 10,11, 12, 13, and 14).

1. Information provided by Hamdallah Zedan, Coordinator, Biodiversity and Biotechnology, United Nations Environment Programme, Nairobi.

Annex C • Criteria for Identifying Countries with Advanced Experience in Strategic Biodiversity Planning

Criteria for Identifying Countries with Advanced Experience in Strategic Biodiversity Planning[1]

PRIMARY CRITERIA

1. Biodiversity Planning is Prescriptive.

The planning process and the document prepared by the country are not merely descriptive; they also provide analysis and prescribe the phases designed to help in strategic decision making. Such plans may be integrated into national development plans, national conservation strategies, or national environment plans, or into plans and strategies on supra- or sub-national level.

2. The Plan is Targeted to Key Institutions.

In the planning process, the strategic analysis and prescriptions (options, activities, policies, investments) are developed in cooperation with the public and private institutions responsible for biodiversity, biological resources, and associated portfolios, and the document is targeted to these decision makers and implementors.

3. Plans are Drawn from all the World Regions.

Cases will be selected to ensure regional balance and to illustrate the peculiar challenges and experiences of developing countries, countries with economies in transition, industrialized nations, and small island states.

SECONDARY CRITERIA

Some of the same cases subjected to the primary criteria will—along with others from the original list—be subjected to these secondary criteria:

4. Plans are Related to Overall National Planning.

At least one case will illustrate a plan developed within the context of integral national planning.

5. Plans Address Funding, Technology, and other Mechanisms Provided for in the Biodiversity Convention.

Cases will be selected that illustrate how countries specifically addressed the articles of the Convention.

6. Plans Illustrate Strategic Biodiversity Planning for Marine and Coastal Ecosystems.

Because these environments are seldom considered fully in national-scale efforts, at least one case will be selected to illustrate experience with these areas and resources at the national or regional scale (e.g., regional seas).

7. Plans Illustrate Linkages to Larger Regional Ecosystems.

At least one case will illustrate how a country incorporated into its strategic analysis the relationship of its territory, responsibilities, and opportunities to the larger multi-country or high-seas regional ecosystem, economic system, and social system of which it is a part.

8. Plans Illustrate Specific Elements of Biodiversity Planning.

Some cases will explicitly illustrate how a country has approached one or more key planning elements—whether to include the setting of goals and objectives, inventory and data management, the evaluation of biotic and institutional assets, plans for national protected area systems, species and genetic resource management plans, ecosystem restoration and species re-introduction, biotechnology, biosafety, capacity building, technology, and funding.

1. The criteria were developed by the World Resources Institute for the purpose of this study.

Annex D • *Pro Forma* Questionnaire for Case Study Reports

Pro Forma Questionnaire for Case Study Reports

REQUESTED INFORMATION FOR COUNTRY CASE STUDIES ON BIODIVERSITY PLANNING

Your interest in participating in this study on biodiversity planning for UNEP is very much appreciated. Your country is well advanced in the process of national biodiversity planning and much can be learned from your experience. Your contribution will help form the information base from which WRI, in cooperation with IUCN, will prepare a report to UNEP and the Biodiversity Convention Secretariat on the approaches being developed by nations as they determine what action they will take to conserve their biodiversity and shift to sustainable resource use through Country Studies, National Strategies, and Action Plans. This document will be published and distributed widely with the aim of sharing the growing body of experience from around the world.

Please prepare a report of no more than 10 pages plus annexes, responding to the following questions in the same order as they appear below.

1. Institutional Base:

Which institution(s) took the lead in preparing the biodiversity plan (Country Study, National Strategy, Action Plan(s), etc.)? If it was a government agency, where in the government does the institution fit, and what is its relationship to other institutions? What was the source of their mandate to take this step?

2. Methodology:

What method, approach, or procedure was followed in the planning effort? What was its origin, e.g., who prepared the method and how was it developed? What groups were involved in designing the method? If your country has prepared a UNEP Biodiversity Country Study, is it being used for the preparation of a National Biodiversity Strategy or Action Plan? Please outline the method as a series of steps, or attach a copy if already available.

3. Participation:

What groups and institutions were invited to participate in the process of preparing the biodiversity plan, and what was the rationale for their involvement? Please characterize the groups, e.g., government, nongovernmental organizations, private sector, grassroots, etc. Did these groups include those expected to be responsible for implementation of the plan? Did these groups participate in all steps of the method (item 2), or only certain steps? Please elaborate. Which groups chose not to be involved, or were excluded? Why? Please provide a list of groups that participated.

4. Goals and Objectives:

What were the overall goals and operational objectives of the plan? How were these established? Who participated in this debate and determination?

5. Development Plan:

Was the biodiversity-planning exercise related to the National Development Plan, a particular national policy, law, or other national process? If so, please describe this relationship. If not, and the country has a development plan or planning body, how will the biodiversity plan be related to the national plan or policy? Please give an example. What other related plans have been prepared that address biodiversity planning (Tropical Forest Action Plan, Country Environmental Profiles, National Environmental Action Plans, UNCED National Reports, Environment Policy and Management Studies, IUCN National Conservation Strategies, UNEP Country Studies, etc.)?

6. Decision Making:

Did the planning process explicitly focus upon providing information and policy guidance for those who will make decisions (executive branch of government, parliament, heads of resource departments, local leaders, NGO directors, business executives, etc.)? How? Please give examples.

7. Biodiversity Convention:

Did the planning method explicitly address the articles of the Convention? How was this done? Please give some specific examples. Now that the Convention has entered into force, how could it be linked more effectively with the Strategy or Action Plan?

8. National, Regional, and Local Scales:

Did the biodiversity-planning exercise address (within country) regional and local issues as well as national? Did it consider ecosystems, species, and genetic materials that are shared with neighboring countries, or relate to CITES, Migratory Species, Ramsar Wetlands, World Heritage, Regional or other Conventions and international agreements? Was there any cooperation with neighboring countries in the planning process? How? Please give specific examples.

9. International Assistance:

Did your biodiversity-planning effort receive technical or financial support from international sources? If so, from whom? Please explain the nature of the relationship.

10. Monitoring and Follow-up:

Was a monitoring process proposed in the plan? Were specific agencies and other institutions and groups assigned particular responsibilities for implementation? How will follow-up action be reported and evaluated? Please provide examples for these aspects of the plan.

11. Obstacles:

Please list the obstacles, impediments, or problems that were encountered in preparing the biodiversity plan, and discuss how these obstacles were faced and overcome. These may be legal, policy, institutional, scientific, technological, or of other origin.

12. Facilitating Factors:

Please list, discuss, and give examples of any factors that you consider greatly facilitated the process of preparing the plan (legal, policy, etc.). If the plan has been completed, to whom has it been submitted (parliament, the executive branch of government, local committees, industry)? What have been the results to date (approval of government, endorsement by local communities, support from industry, budgetary process, executive order, etc.)?

13. Cost:

How long did the process take? How much did it cost? Who financed the process and why? Approximately how many people worked on it?

14. Plan:

Please provide a copy of the biodiversity plan in draft or final form as well as copies of maps and other graphics used to illustrate the plan. Can we employ these materials in the publication, including the maps and graphics, to help illustrate your work and experience? We will, of course, provide copies of our draft report (*National Biodiversity Planning: Guidelines Based on Early Experiences Around the World*) and welcome your critique and further suggestions.

Annex E • Contributors to the Publication

Contributors to the Publication

SECTION 1. COLLABORATING COUNTRIES AND CASE STUDY CONTRIBUTORS

Australia

Josephine Mummery
Biodiversity Unit
Department of the Environment, Sport and Territories
GPO Box 787
Canberra, ACT 2601
Australia
tel: 06-274-1462
fax: 06-274-1895

Canada

John Herity
Biodiversity Convention Office
Environment Canada
351 St. Joseph Blvd.
Hull, Quebec
Canada K1A 0H3
tel: (819) 953-9669
fax: (819) 953-1765

Chile

Consuelo Muñoz
Director, Program in Biodiversity
National Commission on Environment (CONAMA)
Alamenda 949 Piso 13
Santiago, Chile
tel: (56 2) 699-2476 or 699-6127
fax: (56 2) 671-8805

China

Wang Sung
Executive Vice Chairman
Endangered Species Scientific Commission
c/o Institute of Zoology
Chinese Academy of Sciences
19 Zhongguancun Lu
Haidian, Beijing 100080
China
tel/fax: (86 1) 256-2717

Wang Enmin
Research Associate
Institute of Zoology
Chinese Academy of Sciences
19 Zhongguancun Lu
Haidian, Beijing 100080
China
tel: (86 1) 256-2712
fax: (86 1) 256-5689

Costa Rica

Rodrigo Gámez
Director
Instituto Nacional de Biodiversidad (INBio)
3100 Santo Domingo de Heredia
San Jose
Costa Rica
tel: (011 506) 236-7692
fax: (011 506) 236-2816

Egypt

Esam Ahmed Elbadry
Head of Biodiversity Unit
Department of Natural Protectorates
Egyptian Environment Affairs Agency
23 A Ismaeil Mohamed Street
Zamalek, Cairo
Egypt
tel: 340 6777
fax: 340 5962

Germany

Marc Auer
Executive Engineer
Federal Ministry for the Environment, Nature Conservation and Nuclear Safety
Godesberger Allee 90
53175 Bonn
Germany
tel: (49 228) 305-2615
fax: (49 228) 305-2695

Indonesia

Charles Barber
Senior Associate
World Resources Institute
from interviews with
- Suraya Afiff (WALHI),
- Hadi Alikodra (Indonesian Ministry of State for Environment),
- Jerry Bisson (USAID - Indonesia),
- Herman Haeruman (Indonesia National Development Planning Agency),
- Kathy MacKinnon (Environment Department, World Bank; formerly of WWF-Indonesia),
- Agus Purnomo (Pelangi Indonesia Institute),
- Mien Rifai (Herbarium Bogoriense, Indonesian Institute of Sciences),
- Setijati Sastraprajda (Center for Research in Biotechnology, Indonesian Institute of Sciences),
- R.E. Soeriaatmadja (Indonesian Ministry of State for Environment),
- Effendy Sumardja (Indonesian Ministry of Forestry),
- R. Yusuf (Indonesian Ministry of Forestry),
- Arief Yuwono (Indonesian Ministry of State for Environment)

Kenya

Mohamed A. Isahakia
Director/Chief Executive
National Museums of Kenya
P.O. Box 40658
Nairobi, Kenya
tel: (254 2) 742-131/4 or 742-161/4
fax: (254 2) 741-424

Richard Bagine
Centre for Biodiversity
National Museums of Kenya
P.O. Box 40658
Nairobi, Kenya

Mexico

Jorge Soberón
Executive Secretary
National Commission for the Knowledge and Use of Biodiversity
Periferico Sur 4118, 1er Piso
Jardines Del Pedregal
01900 Mexico, D.F.
tel: 652 0706 / 652-0629
fax: 652-0843

Jorge Larson
National Commission for the Knowledge and Use of Biodiversity
Periferico Sur 4118, 1er Piso
Jardines Del Pedregal
01900 Mexico, D.F.
tel: 652 0706 / 652-0629
fax: 652-0834

The Netherlands

Marcel L. Vernooy
Ministry of Agriculture, Nature Management and Fisheries
Department of Nature, Forests, Landscape and Wildlife
Division of International Affairs
P.O. Box. 20402
2599 EK The Hague
The Netherlands
tel: 31-70-379-2934
fax: 31-70-379-3751

Bart Romijin
AIDEnvironment
Donker Curtiusstraat 7-523
1051 JL Amsterdam
The Netherlands
tel: 31-20-686-8111/686-5011
fax: 31-20-686-6251

Norway

Peter Schei
Director General
Directorate for Nature Management
Tungasletta 2
7005 Trondheim
Norway
tel: (47 73) 580 500
fax: (47 73) 915 433

Gudrun Schneider
Head of Division
Ministry of Environment
P.O. Box 8013
Department N-0300
Oslo, Norway
tel: (47 22) 349 090
fax: (47 22) 349 560

Philippines

Charles Barber
Senior Associate
World Resources Institute
from interviews with
- Jose Ampeso (Philippine Department of Foreign Affairs),
- Mary Jean Caleda (Philippine Department of Environment and Natural Resources),
- Corazon Catibog-Sinha (Philippine Department of Environment and Natural Resources),
- Richard Edwards (World-Wide Fund for Nature - Philippines),
- Delfin Ganapin (Foundation for the Philippine Environment),
- Christie Nozawa (Haribon Foundation),
- Rene Salazar (SEARICE)

Poland

Andrzej Weigle
National Foundation for Environmental Protection
Krzywickiego st. 9
02-078 Warsaw, Poland
tel: (48-22) 2514-28
fax: (48-22) 25-21-27

South Pacific Region

Iosefatu Reti
South Pacific Regional Environment Programme
PO Box 240
Apia, Western Samoa
tel: (685) 21 929
fax: (685) 20 231

United Kingdom

Roger Bendall
Head of Species Conservation
European Wildlife Division
Department of Environment
Tollgate House
Houlton Street
Bristol BS2 9DJ
United Kingdom
tel: (44 272) 87 8791
fax: (44 272) 87 8182

Vietnam

Jeffrey McNeely
Chief Conservation Officer
IUCN
Rue de Mauverney 28
1196 Gland
Switzerland
tel: 011 41 22 999-0001
fax: 011 41 22 999-0015

SECTION 2. PEER REVIEW WORKSHOP PARTICIPANTS SEPTEMBER 12-15, 1994, WILLIAMSBURG, VIRGINIA, USA

Marc Auer
Executive Engineer
Federal Ministry for the Environment, Nature Conservation and Nuclear Safety
Germany

Roger Bendall
Head
European Wildlife Division
Department of Environment
United Kingdom

Jeremy Carew-Reid
Director
Conservation Services Division
World Conservation Union (IUCN)
Switzerland

Paul Chabeda
United Nations Environment Programme
Kenya

Feargul Duff
United Nations Environment Programme
Kenya

Brad Fraleigh
Biodiversity Convention Office
Environment Canada
Canada

Mohamed Isahakia
Director/Chief Executive
National Museums of Kenya
Kenya

Hans Jeekel
Head of Strategy
Ministry of Agriculture, Nature Management and Fisheries
The Netherlands

Tim Johnson
World Conservation Monitoring Centre
United Kingdom

Fred Kigenyi
Deputy Commissioner for Forestry
Forest Department
Ministry of Natural Resources
Uganda

Steven Lanou
World Resources Institute
USA

Jorge Larson
National Commission for the Knowledge and Use of Biodiversity
Mexico

Kenton Miller
World Resources Institute
USA

Yakubo Moyini
Moyini & Associates
Uganda

Josephine Mummery
Biodiversity Unit
Department of the Environment, Sport and Territories
Australia

Consuelo Muñoz
Director, Program in Biodiversity
National Commission on Environment (CONAMA)
Chile

Martha Rojas
Biodiversity Programme
World Conservation Union (IUCN)
Switzerland

Peter Schei
Director General
Directorate for Nature Management
Norway

Lisa Sullivan
World Resources Institute
USA

Carlos Valerio
Board of Directors
National Biodiversity Institute
Costa Rica

Wang Sung
Executive Vice Chairman
Endangered Species Scientific Commission
Institute of Zoology, Chinese Academy of Sciences
China

SECTION 3. INDIVIDUALS WHO PROVIDED WRITTEN COMMENT ON THE MANUSCRIPT

Marc Auer
Federal Ministry for the Environment, Nature Conservation and Nuclear Safety
(Germany)

Roger Bendall
Department of the Environment
(United Kingdom)

Walter Arensberg
World Resources Institute
(USA)

Jeremy Carew-Reid
The World Conservation Union (IUCN)
(Switzerland)

Corazon Catibog-Sinha
Department of Environment and Natural Resources
(The Philippines)

Marion Cheatle
United Nations Environment Programme
(Kenya)

Laura Lee Dooley
World Resources Institute
(USA)

Eduardo Fuentes
United Nations Development Programme/GEF
(USA)

Rodrigo Gámez
National Biodiversity Institute
(Costa Rica)

John Herity
Environment Canada
(Canada)

John Hough
United Nations Development Programme/GEF
(USA)

Consuelo Muñoz
National Commission on Environment
(Chile)

A. Rahman
United Nations Environment Programme
(Kenya)

Walter Reid
World Resources Institute
(USA)

Martha Rojas
The World Conservation Union
(Switzerland)

Mukul Sanwal
United Nations Environment Programme
(Kenya)

Setijati Sastraprajda
Indonesian Institute of Sciences
(Indonesia)

Wang Sung
Chinese Academy of Sciences
(China)

Andrzej Weigle
National Foundation for Environmental Protection
(Poland)

Hamdallah Zedan
United Nations Environment Programme
(Kenya)

Annex F • Contents of the Convention on Biodiversity to be Addressed in National Strategies and Action Plans

Contents of the Convention on Biodiversity to be Addressed in National Strategies and Action Plans[1]

ARTICLE 3. PRINCIPLE

States have, in accordance with the Charter of the United Nations and the principles of international law, the sovereign right to exploit their own resources pursuant to their own environmental policies, and the responsibility to ensure that activities within their jurisdiction or control do not cause damage to the environment of other States or of areas beyond the limits of national jurisdiction.

ARTICLE 5. COOPERATION

Each Contracting Party shall, as far as possible and as appropriate, cooperate with other Contracting Parties, directly or, where appropriate, through competent international organizations, in respect of areas beyond national jurisdiction and on other matters of mutual interest, for the conservation and sustainable use of biological diversity.

ARTICLE 6. GENERAL MEASURES FOR CONSERVATION AND SUSTAINABLE USE

Each Contracting Party shall, in accordance with its particular conditions and capabilities:

- Develop national strategies, plans or programmes for the conservation and sustainable use of biological diversity or adapt for this purpose existing strategies, plans or programmes which shall reflect, inter alia, the measures set out in this Convention relevant to the Contracting Party concerned; and
- Integrate, as far as possible and as appropriate, the conservation and sustainable use of biological diversity into relevant sectoral or cross-sectoral plans, programmes and policies.

ARTICLE 7. IDENTIFICATION AND MONITORING

Each Contracting Party shall, as far as possible and as appropriate, in particular for the purposes of Articles 8 to 10:

- Identify components of biological diversity important for its conservation and sustainable use having regard to the indicative list of categories set down in Annex I;
- Monitor, through sampling and other techniques, the components of biological diversity identified pursuant to subparagraph (a) above, paying particular attention to those requiring urgent conservation measures and those which offer the greatest potential for sustainable use;
- Identify processes and categories of activities which have or are likely to have significant adverse impacts on the conservation and sustainable use of biological diversity, and monitor their effects through sampling and other techniques; and
- Maintain and organize, by any mechanism, data derived from identification and monitoring activities pursuant to subparagraphs (a), (b) and (c) above.

ARTICLE 8. *IN-SITU* CONSERVATION

Each Contracting Party shall, as far as possible and as appropriate:

- Establish a system of protected areas or areas

where special measures need to be taken to conserve biological diversity;

- Develop, where necessary, guidelines for the selection, establishment and management of protected areas or areas where special measures need to be taken to conserve biological diversity;
- Regulate or manage biological resources important for the conservation of biological diversity whether within or outside protected areas, with a view to ensuring their conservation and sustainable use;
- Promote the protection of ecosystems, natural habitats and the maintenance of viable populations of species in natural surroundings;
- Promote environmentally sound and sustainable development in areas adjacent to protected areas with a view to furthering protection of these areas;
- Rehabilitate and restore degraded ecosystems and promote the recovery of threatened species, inter alia, through the development and implementation of plans or other management strategies;
- Establish or maintain means to regulate, manage or control the risks associated with the use and release of living modified organisms resulting from biotechnology which are likely to have adverse environmental impacts that could affect the conservation and sustainable use of biological diversity, taking also into account the risks to human health;
- Prevent the introduction of, control or eradicate those alien species which threaten ecosystems, habitats or species;
- Endeavour to provide the conditions needed for compatibility between present uses and the conservation of biological diversity and the sustainable use of its components;
- Subject to its national legislation, respect, preserve and maintain knowledge, innovations and practices of indigenous and local communities embodying traditional lifestyles relevant for the conservation and sustainable use of biological diversity and promote their wider application with the approval and involvement of the holders of such knowledge, innovations and practices and encourage the equitable sharing of the benefits arising from the utilization of such knowledge, innovations and practices;
- Develop or maintain necessary legislation and/or other regulatory provisions for the protection of threatened species and populations;
- Where a significant adverse effect on biological diversity has been determined pursuant to Article 7, regulate or manage the relevant processes and categories of activities; and
- Cooperate in providing financial and other support for *in-situ* conservation outlined in subparagraphs (a) to (l) above, particularly to developing countries.

ARTICLE 9. *EX-SITU* CONSERVATION

Each Contracting Party shall, as far as possible and as appropriate, and predominantly for the purpose of complementing *in-situ* measures:

- Adopt measures for the *ex-situ* conservation of components of biological diversity, preferably in the country of origin of such components;
- Establish and maintain facilities for *ex-situ* conservation of and research on plants, animals and micro-organisms, preferably in the country of origin of genetic resources;
- Adopt measures for the recovery and rehabilitation of threatened species and for their reintroduction into their natural habitats under appropriate conditions;
- Regulate and manage collection of biological resources from natural habitats for *ex-situ* conservation purposes so as not to threaten ecosystems and *in-situ* populations of species, except where special temporary *ex-situ* measures are required under subparagraph (c) above; and
- Cooperate in providing financial and other support for *ex-situ* conservation outlined in subparagraphs (a) to (d) above and in the establishment and maintenance of *ex-situ* conservation facilities in developing countries.

ARTICLE 10. SUSTAINABLE USE OF COMPONENTS OF BIOLOGICAL DIVERSITY

Each Contracting Party shall, as far as possible and as appropriate:

- Integrate consideration of the conservation and sustainable use of biological resources into national decision-making;
- Adopt measures relating to the use of biological resources to avoid or minimize adverse impacts on biological diversity;
- Protect and encourage customary use of biological

resources in accordance with traditional cultural practices that are compatible with conservation or sustainable use requirements;

- Support local populations to develop and implement remedial action in degraded areas where biological diversity has been reduced; and
- Encourage cooperation between its governmental authorities and its private sector in developing methods for sustainable use of biological resources.

ARTICLE 11. INCENTIVE MEASURES

Each Contracting Party shall, as far as possible and as appropriate, adopt economically and socially sound measures that act as incentives for the conservation and sustainable use of components of biological diversity.

ARTICLE 12. RESEARCH AND TRAINING

The Contracting Parties, taking into account the special needs of developing countries, shall:

- Establish and maintain programmes for scientific and technical education and training in measures for the identification, conservation and sustainable use of biological diversity and its components and provide support for such education and training for the specific needs of developing countries;
- Promote and encourage research which contributes to the conservation and sustainable use of biological diversity, particularly in developing countries, inter alia, in accordance with decisions of the Conference of the Parties taken in consequence of recommendations of the Subsidiary Body on Scientific, Technical and Technological Advice; and
- In keeping with the provisions of Articles 16, 18 and 20, promote and cooperate in the use of scientific advances in biological diversity research in developing methods for conservation and sustainable use of biological resources.

ARTICLE 13. PUBLIC EDUCATION AND AWARENESS

The Contracting Parties shall:

- Promote and encourage understanding of the importance of, and the measures required for, the conservation of biological diversity, as well as its propagation through media, and the inclusion of these topics in educational programmes; and
- Cooperate, as appropriate, with other States and international organizations in developing educational and public awareness programmes, with respect to conservation and sustainable use of biological diversity.

ARTICLE 14. IMPACT ASSESSMENT AND MINIMIZING ADVERSE IMPACTS

- Each Contracting Party, as far as possible and as appropriate, shall:
  - Introduce appropriate procedures requiring environmental impact assessment of its proposed projects that are likely to have significant adverse effects on biological diversity with a view to avoiding or minimizing such effects and, where appropriate, allow for public participation in such procedures;
  - Introduce appropriate arrangements to ensure that the environmental consequences of its programmes and policies that are likely to have significant adverse impacts on biological diversity are duly taken into account;
  - Promote, on the basis of reciprocity, notification, exchange of information and consultation on activities under their jurisdiction or control which are likely to significantly affect adversely the biological diversity of other States or areas beyond the limits of national jurisdiction, by encouraging the conclusion of bilateral, regional or multilateral arrangements, as appropriate;
  - In the case of imminent or grave danger or damage, originating under its jurisdiction or control, to biological diversity within the area under jurisdiction of other States or in areas beyond the limits of national jurisdiction, notify immediately the potentially affected States of such danger or damage, as well as initiate action to prevent or minimize such danger or damage; and
  - Promote national arrangements for emergency responses to activities or events, whether caused naturally or otherwise, which present a grave and imminent danger to biological diversity and encourage inter-

national cooperation to supplement such national efforts and, where appropriate and agreed by the States or regional economic integration organizations concerned, to establish joint contingency plans.

- The Conference of the Parties shall examine, on the basis of studies to be carried out, the issue of liability and redress, including restoration and compensation, for damage to biological diversity, except where such liability is a purely internal matter.

ARTICLE 15. ACCESS TO GENETIC RESOURCES

- Recognizing the sovereign rights of States over their natural resources, the authority to determine access to genetic resources rests with the national governments and is subject to national legislation.
- Each Contracting Party shall endeavour to create conditions to facilitate access to genetic resources for environmentally sound uses by other Contracting Parties and not to impose restrictions that run counter to the objectives of this Convention.
- For the purpose of this Convention, the genetic resources being provided by a Contracting Party, as referred to in this Article and Articles 16 and 19, are only those that are provided by Contracting Parties that are countries of origin of such resources or by the Parties that have acquired the genetic resources in accordance with this Convention.
- Access, where granted, shall be on mutually agreed terms and subject to the provisions of this Article.
- Access to genetic resources shall be subject to prior informed consent of the Contracting Party providing such resources, unless otherwise determined by that Party.
- Each Contracting Party shall endeavour to develop and carry out scientific research based on genetic resources provided by other Contracting Parties with the full participation of, and where possible in, such Contracting Parties.
- Each Contracting Party shall take legislative, administrative or policy measures, as appropriate, and in accordance with Articles 16 and 19 and, where necessary, through the financial mechanism established by Articles 20 and 21 with the aim of sharing in a fair and equitable way the results of research and development and the benefits arising from the commercial and other utilization of genetic resources with the Contracting Party providing such resources. Such sharing shall be upon mutually agreed terms.

ARTICLE 16. ACCESS TO AND TRANSFER OF TECHNOLOGY

- Each Contracting Party, recognizing that technology includes biotechnology, and that both access to and transfer of technology among Contracting Parties are essential elements for the attainment of the objectives of this Convention, undertakes subject to the provisions of this Article to provide and/or facilitate access for and transfer to other Contracting Parties of technologies that are relevant to the conservation and sustainable use of biological diversity or make use of genetic resources and do not cause significant damage to the environment.
- Access to and transfer of technology referred to in paragraph 1 above to developing countries shall be provided and/or facilitated under fair and most favourable terms, including on concessional and preferential terms where mutually agreed, and, where necessary, in accordance with the financial mechanism established by Articles 20 and 21. In the case of technology subject to patents and other intellectual property rights, such access and transfer shall be provided on terms which recognize and are consistent with the adequate and effective protection of intellectual property rights. The application of this paragraph shall be consistent with paragraphs 3, 4 and 5 below.
- Each Contracting Party shall take legislative, administrative or policy measures, as appropriate, with the aim that Contracting Parties, in particular those that are developing countries, which provide genetic resources are provided access to and transfer of technology which makes use of those resources, on mutually agreed terms, including technology protected by patents and other intellectual property rights, where necessary, through the provisions of Articles 20 and 21 and in accordance with international law and consistent with paragraphs 4 and 5 below.

- Each Contracting Party shall take legislative, administrative or policy measures, as appropriate, with the aim that the private sector facilitates access to, joint development and transfer of technology referred to in paragraph 1 above for the benefit of both governmental institutions and the private sector of developing countries and in this regard shall abide by the obligations included in paragraphs 1, 2 and 3 above.
- The Contracting Parties, recognizing that patents and other intellectual property rights may have an influence on the implementation of this Convention, shall cooperate in this regard subject to national legislation and international law in order to ensure that such rights are supportive of and do not run counter to its objectives.

ARTICLE 17. EXCHANGE OF INFORMATION

- The Contracting Parties shall facilitate the exchange of information, from all publicly available sources, relevant to the conservation and sustainable use of biological diversity, taking into account the special needs of developing countries.
- Such exchange of information shall include exchange of results of technical, scientific and socio-economic research, as well as information on training and surveying programmes, specialized knowledge, indigenous and traditional knowledge as such and in combination with the technologies referred to in Article 16, paragraph 1. It shall also, where feasible, include repatriation of information.

ARTICLE 18. TECHNICAL AND SCIENTIFIC COOPERATION

- The Contracting Parties shall promote international technical and scientific cooperation in the field of conservation and sustainable use of biological diversity, where necessary, through the appropriate international and national institutions.
- Each Contracting Party shall promote technical and scientific cooperation with other Contracting Parties, in particular developing countries, in implementing this Convention, inter alia, through the development and implementation of national policies. In promoting such cooperation, special attention should be given to the development and strengthening of national capabilities, by means of human resources development and institution building.
- The Conference of the Parties, at its first meeting, shall determine how to establish a clearing-house mechanism to promote and facilitate technical and scientific cooperation.
- The Contracting Parties shall, in accordance with national legislation and policies, encourage and develop methods of cooperation for the development and use of technologies, including indigenous and traditional technologies, in pursuance of the objectives of this Convention. For this purpose, the Contracting Parties shall also promote cooperation in the training of personnel and exchange of experts.
- The Contracting Parties shall, subject to mutual agreement, promote the establishment of joint research programmes and joint ventures for the development of technologies relevant to the objectives of this Convention.

ARTICLE 19. HANDLING OF BIOTECHNOLOGY AND DISTRIBUTION OF ITS BENEFITS

- Each Contracting Party shall take legislative, administrative or policy measures, as appropriate, to provide for the effective participation in biotechnological research activities by those Contracting Parties, especially developing countries, which provide the genetic resources for such research, and where feasible in such Contracting Parties.
- Each Contracting Party shall take all practicable measures to promote and advance priority access on a fair and equitable basis by Contracting Parties, especially developing countries, to the results and benefits arising from biotechnologies based upon genetic resources provided by those Contracting Parties. Such access shall be on mutually agreed terms.
- The Parties shall consider the need for and modalities of a protocol setting out appropriate procedures, including, in particular, advance informed agreement, in the field of the safe transfer, handling and use of any living modified organism resulting from biotechnology that may have adverse effect on the conservation

and sustainable use of biological diversity.

- Each Contracting Party shall, directly or by requiring any natural or legal person under its jurisdiction providing the organisms referred to in paragraph 3 above, provide any available information about the use and safety regulations required by that Contracting Party in handling such organisms, as well as any available information on the potential adverse impact of the specific organisms concerned to the Contracting Party into which those organisms are to be introduced.

ARTICLE 20. FINANCIAL RESOURCES

- Each Contracting Party undertakes to provide, in accordance with its capabilities, financial support and incentives in respect of those national activities which are intended to achieve the objectives of this Convention, in accordance with its national plans, priorities and programmes.
- The developed country Parties shall provide new and additional financial resources to enable developing country Parties to meet the agreed full incremental costs to them of implementing measures which fulfil the obligations of this Convention and to benefit from its provisions and which costs are agreed between a developing country Party and the institutional structure referred to in Article 21, in accordance with policy, strategy, programme priorities and eligibility criteria and an indicative list of incremental costs established by the Conference of the Parties. Other Parties, including countries undergoing the process of transition to a market economy, may voluntarily assume the obligations of the developed country Parties. For the purpose of this Article, the Conference of the Parties, shall at its first meeting establish a list of developed country Parties and other Parties which voluntarily assume the obligations of the developed country Parties. The Conference of the Parties shall periodically review and if necessary amend the list. Contributions from other countries and sources on a voluntary basis would also be encouraged. The implementation of these commitments shall take into account the need for adequacy, predictability and timely flow of funds and the importance of burdensharing among the contributing Parties included in the list.
- The developed country Parties may also provide, and developing country Parties avail themselves of, financial resources related to the implementation of this Convention through bilateral, regional and other multilateral channels.
- The extent to which developing country Parties will effectively implement their commitments under this Convention will depend on the effective implementation by developed country Parties of their commitments under this Convention related to financial resources and transfer of technology and will take fully into account the fact that economic and social development and eradication of poverty are the first and overriding priorities of the developing country Parties.
- The Parties shall take full account of the specific needs and special situation of least developed countries in their actions with regard to funding and transfer of technology.
- The Contracting Parties shall also take into consideration the special conditions resulting from the dependence on, distribution and location of, biological diversity within developing country Parties, in particular small island States.
- Consideration shall also be given to the special situation of developing countries, including those that are most environmentally vulnerable, such as those with arid and semi-arid zones, coastal and mountainous areas.

ARTICLE 21. FINANCIAL MECHANISM

- There shall be a mechanism for the provision of financial resources to developing country Parties for purposes of this Convention on a grant or concessional basis the essential elements of which are described in this Article. The mechanism shall function under the authority and guidance of, and be accountable to, the Conference of the Parties for purposes of this Convention. The operations of the mechanism shall be carried out by such institutional structure as may be decided upon by the Conference of the Parties at its first meeting. For purposes of this Convention, the Conference of the Parties shall determine the policy, strategy, programme priorities and eligibility criteria relating to the access to and utilization of such resources. The contributions shall be such as to take into account the need for predictability, adequacy and timely flow of funds referred to in Article 20 in accordance with the amount of resources needed to be decided periodically by the Conference of the Parties and the importance of burdensharing among the contributing Parties included in the list referred to in Article 20, paragraph 2. Voluntary contributions may also be made by the developed country Parties and by other countries and sources. The mechanism shall operate within a democratic and transparent system of governance.

- Pursuant to the objectives of this Convention, the Conference of the Parties shall at its first meeting determine the policy, strategy and programme priorities, as well as detailed criteria and guidelines for eligibility for access to and utilization of the financial resources including monitoring and evaluation on a regular basis of such utilization. The Conference of the Parties shall decide on the arrangements to give effect to paragraph 1 above after consultation with the institutional structure entrusted with the operation of the financial mechanism.
- The Conference of the Parties shall review the effectiveness of the mechanism established under this Article, including the criteria and guidelines referred to in paragraph 2 above, not less than two years after the entry into force of this Convention and thereafter on a regular basis. Based on such review, it shall take appropriate action to improve the effectiveness of the mechanism if necessary.
- The Contracting Parties shall consider strengthening existing financial institutions to provide financial resources for the conservation and sustainable use of biological diversity.

ARTICLE 26. REPORTS

Each Contracting Party shall, at intervals to be determined by the Conference of the Parties, present to the Conference of the Parties, reports on measures which it has taken for the implementation of the provisions of this Convention and their effectiveness in meeting the objectives of this Convention.

1. This is the verbatim text of the Convention on Biological Diversity.

References

ANZECC (Australian and New Zealand Environment and Conservation Council). Undated Draft. *National Strategy for the Conservation of Australia's Biological Diversity.* ANZECC.

BAPPT (Biodiversity Action Plan Planning Team). 1993 Draft. *Biodiversity Action Plan for Vietnam.* Biodiversity Action Plan Planning Team, Hanoi.

Carew-Reid, Jeremy, Robert Prescott-Allen, Stephen Bass and Barry Dalal-Clayton. 1994. *Strategies for National Sustainable Development: A Handbook for their Planning and Implementation.* IIED and IUCN, London and Gland, respectively.

CONAMA (Comision Nacional Del Medio Ambiente) Secretaria Tecnica y Administrativa. 1993. *Propuesta De Plan de Accion Nacional Para La Biodiversidad en Chile.* CONAMA, Santiago.

Department of Environment. 1994a. *Biodiversity: The UK Action Plan.* HMSO, London.

Department of Environment. 1994b. *Biodiversity: The UK Action Plan, Summary Report.* HMSO, London.

DNM (Directorate for Nature Management). 1992. *Biological Diversity in Norway: A Country Study.* DN-Report 1992-5b. DNM, Trondheim.

——.1994. *National Action Plan for Biological Diversity—Guidelines for Sector Plans.* Directorate for Nature Management, Trondheim.

FAO (Food and Agriculture Organization of the United Nations). 1991. *Strategy and Agenda for Action for Sustainable Agriculture and Rural Development.* FAO, Rome.

FAO/WRI/IBRD/UNDP (Food and Agriculture Organization of the United Nations/World Resources Institute/United Nations Development Programme). 1987. *The Tropical Forestry Action Plan.* FAO, Rome.

FPT (Federal-Provincial-Territorial) Biodiversity Working Group. 1994. *Draft Canadian Biodiversity Strategy. Biodiversity Convention Office,* Hull, Quebec.

GEF (Global Environment Facility). 1992. *Criteria for Eligibility and Priorities for Selection for Global Environment Facility Projects.* Global Environment Facility, Washington DC.

Glowka, Lyle, Francoise Burhenne-Guilmin, Hugh Synge, Jeffrey McNeely, and Lothar Gundling. 1994. *A Guide to the Convention on Biological Diversity.* World Conservation Union (IUCN), Cambridge.

GOI (Government of Indonesia). 1992. *Indonesian Country Study on Biological Diversity.* Ministry of State for Population and Environment, Jakarta.

——. 1993a. *Biodiversity Action Plan for Indonesia.* Ministry of National Development Planning/National Development Planning Agency, Jakarta.

——. 1993b. *Biodiversity Action Plan for Indonesia: Project Profile (Selected), Priority Project Repelita VI.* Ministry of National Development Planning/National Development Planning Agency, Jakarta.

——. 1993c. *Strategi Nasional Pengeloaan Keanekaragaman Hayati (National Biodiversity Management Strategy)* Ministry of State for Population and Environment, Jakarta.

IMO (International Maritime Organization). 1983. *Strategy for the Protection of the Marine Environment.* IMO.

IUCN (World Conservation Union). Draft (a). *Strategies for Sustainability: Africa, Vol. 1.* World Conservation Union, Gland.

——. Draft (b). *Strategies for Sustainability: Asia, Vol. 1.* World Conservation Union, Gland.

——. 1994a. *United Nations List of National Parks and Protected Areas.* Compiled by the World Conservation Monitoring Centre and the IUCN Commission on National Parks and Protected Areas. The World Conservation Union, Gland and Cambridge.

——. 1994b. *National Environmental Funds:* A New Mechanism for Conservation Finance. Compiled by Mark Dillenbeck. The World Conservation Union-US, Washington, DC.

——. 1994c. *Goal and Purpose: Global Initiative for National Environmental Funds.* The World Conservation Union-US, Washington, DC.

IUCN, MacArthur Foundation, C.S. Mott Foundation, The Nature Conservancy, UNDP and WWF-US. 1994. *Summary Report on the Global Forum on Environmental Funds.* May 30-June 2, 1994, Santa Cruz, Bolivia.

IUCN/UNEP/WWF (World Conservation Union/United Nations Environment Programme/World Wildlife Fund). 1980. *World Conservation Strategy. Living resource conservation for sustainable development.* IUCN, UNEP and WWF, Gland, Nairobi, Washington, respectively.

IUCN/UNEP/WWF. 1991. *Caring for the Earth. A Strategy for Sustainable Living.* Gland.

Keating, Michael. 1993. *Agenda for Change: A plain language version of Agenda 21 and the other Rio Agreements.* Centre for Our Common Future, Geneva.

Lynch, Owen. 1992. "Securing Community-Based Tenurial Rights in the Tropical Forests of Asia: An Overview of Current and Prospective Strategies." *Issues in Development November 1992.* World Resources Institute (WRI), Washington D.C.

Mace, Georgina and Simon Stuart. 1994. *Draft IUCN Red List Category: Species Vol. 22.* The World Conservation Union, Gland. Pp. 13-24.

MacKinnon, John. 1994. "Too Many Plans: views on the planning process." Paper presented at the IUCN General Assembly, Workshop on Biodiversity, January 1994. Buenos Aires.

McNeely, Jeffrey A. 1988. *Economics and Biological Diversity: Developing and Using Economic Incentives to Conserve Biological Diversity.* The World Conservation Union, Gland.

McNeely, Jeffrey A., Kenton R. Miller, Walter V. Reid, Russell A. Mittermeier and Timothy B. Werner. 1990. *Conserving the World's Biological Diversity.* The World Conservation Union, World Resources Institute, Conservation International, World Wide Fund for Nature US and The World Bank, Gland and Washington D.C.

Miller, Kenton R., 1995. *Balancing the Scales: Policies for increasing biodiversity's chances through bioregional management.* World Resources Institute, Washington D.C. (Draft manuscript).

MIRENEM/MNCR/INBio. 1992. *Estudio Nacional de Biodiversidad: Costos, Beneficios y Necesidades de Financiamiento de la Conservación de la Diversidad Biológica en Costa Rica.* Gobierno de Costa Rica and Ministerio de Recursos Naturales, Energia y Minas, San José.

NEPA (National Environmental Protection Agency). 1994. *China Biodiversity Conservation Action Plan.* National Environmental Protection Agency, Beijing.

Protected Areas and Wildlife Bureau, DENR and Sub-Committee on Biodiversity, Philippine Council for Sustainable Development. Undated Draft. *Philippine Strategy for Biological Diversity Conservation (PSBDC).* Sub-Committee on Biological Diversity of the Philippine Council for Sustainable Development, Manila.

Reid, Walter V. and Kenton R. Miller. 1989. *Keeping Options Alive: The Scientific Basis for Conserving Biodiversity.* World Resources Institute, Washington D.C.

Reid, Walter V., Jeffrey A. McNeely, Daniel B. Tunstall, Dirk A. Bryant and Manuel Winograd. 1993a. *Biodiversity Indicators for Policy-Makers.* World Resources Institute, Washington D.C.

Reid, Walter V., Sarah A. Laird, Carrie A Meyer, Rodrigo Gámez, Ana Sittenfeld, Daniel H. Janzen, Michael A. Gollin and Calestous Juma. 1993b. *Biodiversity Prospecting: Using Genetic Resources for Sustainable Development.* World Resources Institute, Washington D.C.

Therivel, Riki, Elizabeth Wilson, Stewart Thompson, Donna Heaney and David Pritchard. 1992. *Strategic Environmental Assessment.* Earthscan Publications Ltd., London.

UN (United Nations). 1993. *Agenda 21, Rio Declaration, Forest Principles: The Final Text of Agreements.* United Nations, New York.

UNCHS (United Nations Conference on Human Settlements). 1976. *Vancouver Action Plan for Human Settlements.* Adopted by UNCHS, Vancouver.

UNCOD (United Nations Conference on Desertification). 1977. *Plan of Action to Combat Desertification.* Adopted by UNCOD, Nairobi.

UNDP/IADB (United Nations Development Programme/Inter-American Development Bank). 1990. *Our Own Agenda: Report of the Latin American and Caribbean Commission on Development and Environment.* UNDP, New York.

UNEP (United Nations Environment Programme). 1992. *Convention on Biological Diversity.* UNEP, Nairobi.

——. 1993a. *Guidelines for Country Studies on Biological Diversity.* UNEP, Nairobi.

——. 1993b. *Report of Panel I: Priorities for Action for Conservation and Sustainable Use of Biological Diversity and Agenda for Scientific and Technological Research.* A Report of the Expert Panels Established to Follow-up on the Convention on Biological Diversity. UNEP, Nairobi.

——. 1993c. *Report of Panel II: Evaluation of Economic Implications of Conservation of Biological Diversity and Its Sustainable Use and Evaluation of Biological and Genetic Resources.* A report of the Expert Panels Established to Follow-up on the Convention on Biological Diversity. UNEP, Nairobi.

——. 1993d. *Technology Transfer and Financial Issues: Issues and Options from Panel III.* A report of the Expert Panels Established to Follow-up on the Convention on Biological Diversity. UNEP, Nairobi.

UNEP/UNESCO (United Nations Environment Programme/United Nations Educational, Scientific, and Cultural Organization) 1975. *International Environmental Education Programme.* Nairobi, Paris.

UNWC (United Nations Water Conference). 1977. *Mar del Plata Action Plan for Water Resources Development.* Adopted by the UNWC, Mar del Plata, Argentina.

WCED (World Commission on Environment and Development). 1987. *Our Common Future.* Oxford University Press, Oxford.

Wilson, David (ed.) 1990. *Environmental Impact Assessment: Theory and Practice.* Routledge, London and New York.

World Bank. 1991a. *Environmental Assessment Sourcebook Volume I: Policies, Procedures, and Cross-Sectoral Issues.* World Bank Technical Paper Number 139. The World Bank, Washington, D.C.

——. 1991b. *Environmental Assessment Sourcebook Volume II: Sectoral Guidelines.* World Bank Technical Paper Number 140. The World Bank, Washington, D.C.

——. 1991c. *Environmental Assessment Sourcebook Volume III: Guidelines for Environmental Assessment of Energy and Industry Projects.* World Bank Technical Paper Number 154. The World Bank, Washington, D.C.

WCMC (World Conservation Monitoring Centre). 1992. *Global Biodiversity: Status of the Earth's Living Resources.* Chapman and Hall, London.

——. 1994a. B. Groombridge (ed.) *Biodiversity Data Sourcebook.* World Conservation Press, Cambridge, UK.

——. 1994b. *Priorities for Conserving Global Species Richness and Endemism.* WCMC, Cambridge.

WHO (World Health Organization). 1981. *Global Strategy for Health for All by the Year 2000.* World Health Organization, Geneva.

WPC (World Population Conference). 1974. *World Population Plan of Action.* Adopted by WPC, Bucharest.

WRI (World Resources Institute). 1992. *Public Policy and Legislation in Environmental Management: Terracing in Nyarurembo, Uganda.* From the Ground Up Case Study No. 5. World Resources Institute, Washington, D.C.

WRI/IIED/IUCN. Daniel Tunstall and Mieke van der Wansem (eds.) 1992. *1993 Directory of Country Environmental Studies.* World Resources Institute, Washington, D.C.

WRI/IUCN/UNEP. (World Resources Institute/World Conservation Union/United Nations Environment Programme). 1992. *Global Biodiversity Strategy: Guidelines for Action to Save, Study, and Use Earth's Biotic Wealth Sustainably and Equitably.* WRI/IUCN/UNEP, Washington D.C, Gland, Switzerland, and Nairobi.